T0207281

Lecture Notes in Computer Science 14733

The series Lecture Notes in Computer Science (LNCS), including its subseries Lecture Notes in Artificial Intelligence (LNAI) and Lecture Notes in Bioinformatics (LNBI), has established itself as a medium for the publication of new developments in computer science and information technology research, teaching, and education.

LNCS enjoys close cooperation with the computer science R & D community, the series counts many renowned academics among its volume editors and paper authors, and collaborates with prestigious societies. Its mission is to serve this international community by providing an invaluable service, mainly focused on the publication of conference and workshop proceedings and postproceedings. LNCS commenced publication in 1973.

Heidi Krömker
Editor

HCI in Mobility, Transport, and Automotive Systems

6th International Conference, MobiTAS 2024
Held as Part of the 26th HCI International Conference, HCII 2024
Washington, DC, USA, June 29 – July 4, 2024
Proceedings, Part II

Editor
Heidi Krömker
Technische Universitat Ilmenau
Ilmenau, Germany

ISSN 0302-9743 ISSN 1611-3349 (electronic)
Lecture Notes in Computer Science
ISBN 978-3-031-60479-9 ISBN 978-3-031-60480-5 (eBook)
https://doi.org/10.1007/978-3-031-60480-5

This Springer imprint is published by the registered company Springer Nature Switzerland AG
The registered company address is: Gewerbestrasse 11, 6330 Cham, Switzerland

Foreword

This year we celebrate 40 years since the establishment of the HCI International (HCII) Conference, which has been a hub for presenting groundbreaking research and novel ideas and collaboration for people from all over the world.

The HCII conference was founded in 1984 by Prof. Gavriel Salvendy (Purdue University, USA, Tsinghua University, P.R. China, and University of Central Florida, USA) and the first event of the series, "1st USA-Japan Conference on Human-Computer Interaction", was held in Honolulu, Hawaii, USA, 18–20 August. Since then, HCI International is held jointly with several Thematic Areas and Affiliated Conferences, with each one under the auspices of a distinguished international Program Board and under one management and one registration. Twenty-six HCI International Conferences have been organized so far (every two years until 2013, and annually thereafter).

Over the years, this conference has served as a platform for scholars, researchers, industry experts and students to exchange ideas, connect, and address challenges in the ever-evolving HCI field. Throughout these 40 years, the conference has evolved itself, adapting to new technologies and emerging trends, while staying committed to its core mission of advancing knowledge and driving change.

As we celebrate this milestone anniversary, we reflect on the contributions of its founding members and appreciate the commitment of its current and past Affiliated Conference Program Board Chairs and members. We are also thankful to all past conference attendees who have shaped this community into what it is today.

The 26th International Conference on Human-Computer Interaction, HCI International 2024 (HCII 2024), was held as a 'hybrid' event at the Washington Hilton Hotel, Washington, DC, USA, during 29 June – 4 July 2024. It incorporated the 21 thematic areas and affiliated conferences listed below.

A total of 5108 individuals from academia, research institutes, industry, and government agencies from 85 countries submitted contributions, and 1271 papers and 309 posters were included in the volumes of the proceedings that were published just before the start of the conference, these are listed below. The contributions thoroughly cover the entire field of human-computer interaction, addressing major advances in knowledge and effective use of computers in a variety of application areas. These papers provide academics, researchers, engineers, scientists, practitioners and students with state-of-the-art information on the most recent advances in HCI.

The HCI International (HCII) conference also offers the option of presenting 'Late Breaking Work', and this applies both for papers and posters, with corresponding volumes of proceedings that will be published after the conference. Full papers will be included in the 'HCII 2024 - Late Breaking Papers' volumes of the proceedings to be published in the Springer LNCS series, while 'Poster Extended Abstracts' will be included as short research papers in the 'HCII 2024 - Late Breaking Posters' volumes to be published in the Springer CCIS series.

I would like to thank the Program Board Chairs and the members of the Program Boards of all thematic areas and affiliated conferences for their contribution towards the high scientific quality and overall success of the HCI International 2024 conference. Their manifold support in terms of paper reviewing (single-blind review process, with a minimum of two reviews per submission), session organization and their willingness to act as goodwill ambassadors for the conference is most highly appreciated.

This conference would not have been possible without the continuous and unwavering support and advice of Gavriel Salvendy, founder, General Chair Emeritus, and Scientific Advisor. For his outstanding efforts, I would like to express my sincere appreciation to Abbas Moallem, Communications Chair and Editor of HCI International News.

July 2024 Constantine Stephanidis

HCI International 2024 Thematic Areas and Affiliated Conferences

- HCI: Human-Computer Interaction Thematic Area
- HIMI: Human Interface and the Management of Information Thematic Area
- EPCE: 21st International Conference on Engineering Psychology and Cognitive Ergonomics
- AC: 18th International Conference on Augmented Cognition
- UAHCI: 18th International Conference on Universal Access in Human-Computer Interaction
- CCD: 16th International Conference on Cross-Cultural Design
- SCSM: 16th International Conference on Social Computing and Social Media
- VAMR: 16th International Conference on Virtual, Augmented and Mixed Reality
- DHM: 15th International Conference on Digital Human Modeling & Applications in Health, Safety, Ergonomics & Risk Management
- DUXU: 13th International Conference on Design, User Experience and Usability
- C&C: 12th International Conference on Culture and Computing
- DAPI: 12th International Conference on Distributed, Ambient and Pervasive Interactions
- HCIBGO: 11th International Conference on HCI in Business, Government and Organizations
- LCT: 11th International Conference on Learning and Collaboration Technologies
- ITAP: 10th International Conference on Human Aspects of IT for the Aged Population
- AIS: 6th International Conference on Adaptive Instructional Systems
- HCI-CPT: 6th International Conference on HCI for Cybersecurity, Privacy and Trust
- HCI-Games: 6th International Conference on HCI in Games
- MobiTAS: 6th International Conference on HCI in Mobility, Transport and Automotive Systems
- AI-HCI: 5th International Conference on Artificial Intelligence in HCI
- MOBILE: 5th International Conference on Human-Centered Design, Operation and Evaluation of Mobile Communications

List of Conference Proceedings Volumes Appearing Before the Conference

1. LNCS 14684, Human-Computer Interaction: Part I, edited by Masaaki Kurosu and Ayako Hashizume
2. LNCS 14685, Human-Computer Interaction: Part II, edited by Masaaki Kurosu and Ayako Hashizume
3. LNCS 14686, Human-Computer Interaction: Part III, edited by Masaaki Kurosu and Ayako Hashizume
4. LNCS 14687, Human-Computer Interaction: Part IV, edited by Masaaki Kurosu and Ayako Hashizume
5. LNCS 14688, Human-Computer Interaction: Part V, edited by Masaaki Kurosu and Ayako Hashizume
6. LNCS 14689, Human Interface and the Management of Information: Part I, edited by Hirohiko Mori and Yumi Asahi
7. LNCS 14690, Human Interface and the Management of Information: Part II, edited by Hirohiko Mori and Yumi Asahi
8. LNCS 14691, Human Interface and the Management of Information: Part III, edited by Hirohiko Mori and Yumi Asahi
9. LNAI 14692, Engineering Psychology and Cognitive Ergonomics: Part I, edited by Don Harris and Wen-Chin Li
10. LNAI 14693, Engineering Psychology and Cognitive Ergonomics: Part II, edited by Don Harris and Wen-Chin Li
11. LNAI 14694, Augmented Cognition, Part I, edited by Dylan D. Schmorrow and Cali M. Fidopiastis
12. LNAI 14695, Augmented Cognition, Part II, edited by Dylan D. Schmorrow and Cali M. Fidopiastis
13. LNCS 14696, Universal Access in Human-Computer Interaction: Part I, edited by Margherita Antona and Constantine Stephanidis
14. LNCS 14697, Universal Access in Human-Computer Interaction: Part II, edited by Margherita Antona and Constantine Stephanidis
15. LNCS 14698, Universal Access in Human-Computer Interaction: Part III, edited by Margherita Antona and Constantine Stephanidis
16. LNCS 14699, Cross-Cultural Design: Part I, edited by Pei-Luen Patrick Rau
17. LNCS 14700, Cross-Cultural Design: Part II, edited by Pei-Luen Patrick Rau
18. LNCS 14701, Cross-Cultural Design: Part III, edited by Pei-Luen Patrick Rau
19. LNCS 14702, Cross-Cultural Design: Part IV, edited by Pei-Luen Patrick Rau
20. LNCS 14703, Social Computing and Social Media: Part I, edited by Adela Coman and Simona Vasilache
21. LNCS 14704, Social Computing and Social Media: Part II, edited by Adela Coman and Simona Vasilache
22. LNCS 14705, Social Computing and Social Media: Part III, edited by Adela Coman and Simona Vasilache

23. LNCS 14706, Virtual, Augmented and Mixed Reality: Part I, edited by Jessie Y. C. Chen and Gino Fragomeni
24. LNCS 14707, Virtual, Augmented and Mixed Reality: Part II, edited by Jessie Y. C. Chen and Gino Fragomeni
25. LNCS 14708, Virtual, Augmented and Mixed Reality: Part III, edited by Jessie Y. C. Chen and Gino Fragomeni
26. LNCS 14709, Digital Human Modeling and Applications in Health, Safety, Ergonomics and Risk Management: Part I, edited by Vincent G. Duffy
27. LNCS 14710, Digital Human Modeling and Applications in Health, Safety, Ergonomics and Risk Management: Part II, edited by Vincent G. Duffy
28. LNCS 14711, Digital Human Modeling and Applications in Health, Safety, Ergonomics and Risk Management: Part III, edited by Vincent G. Duffy
29. LNCS 14712, Design, User Experience, and Usability: Part I, edited by Aaron Marcus, Elizabeth Rosenzweig and Marcelo M. Soares
30. LNCS 14713, Design, User Experience, and Usability: Part II, edited by Aaron Marcus, Elizabeth Rosenzweig and Marcelo M. Soares
31. LNCS 14714, Design, User Experience, and Usability: Part III, edited by Aaron Marcus, Elizabeth Rosenzweig and Marcelo M. Soares
32. LNCS 14715, Design, User Experience, and Usability: Part IV, edited by Aaron Marcus, Elizabeth Rosenzweig and Marcelo M. Soares
33. LNCS 14716, Design, User Experience, and Usability: Part V, edited by Aaron Marcus, Elizabeth Rosenzweig and Marcelo M. Soares
34. LNCS 14717, Culture and Computing, edited by Matthias Rauterberg
35. LNCS 14718, Distributed, Ambient and Pervasive Interactions: Part I, edited by Norbert A. Streitz and Shin'ichi Konomi
36. LNCS 14719, Distributed, Ambient and Pervasive Interactions: Part II, edited by Norbert A. Streitz and Shin'ichi Konomi
37. LNCS 14720, HCI in Business, Government and Organizations: Part I, edited by Fiona Fui-Hoon Nah and Keng Leng Siau
38. LNCS 14721, HCI in Business, Government and Organizations: Part II, edited by Fiona Fui-Hoon Nah and Keng Leng Siau
39. LNCS 14722, Learning and Collaboration Technologies: Part I, edited by Panayiotis Zaphiris and Andri Ioannou
40. LNCS 14723, Learning and Collaboration Technologies: Part II, edited by Panayiotis Zaphiris and Andri Ioannou
41. LNCS 14724, Learning and Collaboration Technologies: Part III, edited by Panayiotis Zaphiris and Andri Ioannou
42. LNCS 14725, Human Aspects of IT for the Aged Population: Part I, edited by Qin Gao and Jia Zhou
43. LNCS 14726, Human Aspects of IT for the Aged Population: Part II, edited by Qin Gao and Jia Zhou
44. LNCS 14727, Adaptive Instructional System, edited by Robert A. Sottilare and Jessica Schwarz
45. LNCS 14728, HCI for Cybersecurity, Privacy and Trust: Part I, edited by Abbas Moallem
46. LNCS 14729, HCI for Cybersecurity, Privacy and Trust: Part II, edited by Abbas Moallem

https://2024.hci.international/proceedings

https://2024.hci.international/proceedings

Preface

Human-computer interaction in the highly complex field of mobility and intermodal transport leads to completely new challenges. A variety of different travelers move in different travel chains. The interplay of such different systems, such as car and bike sharing, local and long-distance public transport, and individual transport, must be adapted to the needs of travelers. Intelligent traveler information systems must be created to make it easier for travelers to plan, book, and execute an intermodal travel chain and to interact with the different systems. Innovative means of transport are developed, such as electric vehicles and autonomous vehicles. To achieve the acceptance of these systems, human-machine interaction must be completely redesigned.

The 6th International Conference on HCI in Mobility, Transport, and Automotive Systems (MobiTAS 2024), an affiliated conference of the HCI International (HCII) conference, encouraged papers from academics, researchers, industry, and professionals, on a broad range of theoretical and applied issues related to mobility, transport, and automotive systems and their applications.

For MobiTAS 2024, a key theme with which researchers were concerned was the safety and well-being of drivers. From investigating the correlation between motion sickness and driving activities to exploring the effects of advanced head-up display technologies on driver behavior, the effects of alerts, take over strategies, as well as driver behavior and performance, each contribution offers valuable insights into the complexities of human factors in driving. A considerable number of submissions focused on the multifaceted dynamics of human trust, emotion, and cognition in the context of automated driving, exploring also autonomous driving scenarios, connected vehicles and teleoperation systems, offering invaluable insights into the future of transportation. Furthermore, in this era of rapid technological advancements and evolving user needs, several papers focused on enhanced inclusivity and user experience, discussing inclusive mobility and assistive systems, independent travel for people with intellectual disabilities, cognitive load in automotive interfaces, enhanced interaction through gestures, as well as user experience design and usability testing approaches. Finally, the design of urban mobility and public transportation systems has also collected diverse contributions on user needs and perspectives regarding public transportation, as well as human-centered design and user acceptance of urban mobility approaches, contributing to the ongoing dialogue for innovations in public transportation.

Two volumes of the HCII 2024 proceedings are dedicated to this year's edition of the MobiTAS conference. The first focuses on topics related to Driver Behavior and Safety, and Human Factors in Automated Vehicles, while the second focuses on topics related to Urban Mobility and Public Transportation, and User Experience and Inclusivity in MobiTAS.

The papers in these volumes were accepted for publication after a minimum of two single-blind reviews from the members of the MobiTAS Program Board or, in some cases, from members of the Program Boards of other affiliated conferences. I would like to thank all of them for their invaluable contribution, support, and efforts.

July 2024 Heidi Krömker

6th International Conference on HCI in Mobility, Transport and Automotive Systems (MobiTAS 2024)

The full list with the Program Board Chairs and the members of the Program Boards of all thematic areas and affiliated conferences of HCII 2024 is available online at:

http://www.hci.international/board-members-2024.php

HCI International 2025 Conference

The 27th International Conference on Human-Computer Interaction, HCI International 2025, will be held jointly with the affiliated conferences at the Swedish Exhibition & Congress Centre and Gothia Towers Hotel, Gothenburg, Sweden, June 22–27, 2025. It will cover a broad spectrum of themes related to Human-Computer Interaction, including theoretical issues, methods, tools, processes, and case studies in HCI design, as well as novel interaction techniques, interfaces, and applications. The proceedings will be published by Springer. More information will become available on the conference website: https://2025.hci.international/.

General Chair
Prof. Constantine Stephanidis
University of Crete and ICS-FORTH
Heraklion, Crete, Greece
Email: general_chair@2025.hci.international

https://2025.hci.international/

Contents – Part II

Contents – Part I

Driver Behavior and Safety

Human Factors in Automated Vehicles

Urban Mobility and Public Transportation

Next Stop: Passenger Perspectives on Autonomous Trains

Andrea Arzer(✉), Lauren Beehler(✉), and Marloes Vredenborg

Utrecht University, Utrecht, The Netherlands
{a.arzer,l.a.beehler}@students.uu.nl, m.t.r.vredenborg@uu.nl

Abstract. While extensive research exists on autonomous cars for private use, there is a notable gap in understanding public opinions on autonomous trains. Understanding passengers' views on a technology that might be available to them in the next decade could highly influence its success. This paper researches factors influencing potential passengers when deciding to ride fully autonomous trains and explores solutions to counteract negative perceptions. To complement the available research on this topic that has been conducted using quantitative methods, this paper describes a multi-method qualitative study combining focus groups and creative problem-solving sessions. Key findings include participants' distrust in unfamiliar systems and hesitation about the absence of human staff onboard. Proposed ideas include the visible implementation of additional safety features available to the passengers, the adaptation of the train interior to make it more inviting, and the provision of information about the operation of the autonomous trains. This study uncovered different perspectives and concerns related to autonomous railway vehicles, along with solutions that can be implemented to increase passenger trust. However, it also emphasizes the complexity of the topic, illustrating the necessity for additional research.

Keywords: Autonomous vehicle · Train · Technology acceptance · Human-centered perspective · Trust · Qualitative research methods

1 Introduction

Trains are the most-used public transportation modality in the Netherlands; on average, each resident traveled 610 km by rail in 2021 [42]. However, difficulties are expected to arise from the combination of the predicted increase in train usage and existing staff shortages [44]. Autonomous trains (ATs) may represent a new solution to overcome these and other problems. ATs are more efficient and safer, due in part to the elimination of human error [14,40]. ATs can be controlled more precisely, allowing for increased service frequency, reliability, and flexibility of train service, and reduced passenger waiting time [40]. The implementation of ATs is not a new idea in Europe: Germany is studying and

A. Arzer and L. Beehler—The two authors contributed equally to this paper.

H. Krömker (Ed.): HCII 2024, LNCS 14733, pp. 3–25, 2024.
https://doi.org/10.1007/978-3-031-60480-5_1

testing the implementation of fully-automated trains for public transport [5,24], and France's and Italy's use constitutes most of the automated systems in the European Union [22,23,48]. Given the advantages and successful examples set by other countries, it is not surprising that the Netherlands is considering the implementation of passenger ATs [2].

Implementation, however, may bring additional challenges, evidenced by the response to autonomous road vehicles (e.g., cars and buses). While many cars already have elements of automation, like the well-known driver assistance, fully autonomous cars are currently undergoing testing [8,43]. Although autonomous cars might offer increased safety and increased free time, they are regarded critically by the public. Literature on autonomous cars features contrasting opinions regarding usefulness, convenience, and costs [20,33,37], but user safety and data protection are more unanimous concerns [6,25,27,38]. Moreover, news and statistics about crashes involving autonomous vehicles are judged and recalled more harshly than crashes of human-driven vehicles, likely leading to reduced trust [31]. Yet, other research reports that people expect autonomous cars to drive more conservatively than human-driven vehicles, increasing their trustworthiness [10]. People seem to have a more positive attitude about autonomous buses [4] and are willing to use them in the future [13,34]. Still, when exploring efficiency, effectiveness, usefulness, and performance expectancy, people express concerns [12,34,47]. Personal experience and demographic and socioeconomic factors can influence perceptions of autonomous buses [1,13,30,36].

This pushback indicates that people could also develop concerns with ATs with more exposure. Since social acceptance is a crucial factor in the success of new technologies, public concerns might prevent AT implementation or success. Limited research has studied people's perceptions of ATs, which operate in markedly different contexts than autonomous road vehicles (i.e., fixed tracks and less interaction with other vehicles). Some studies used quantitative methods to analyze people's perspectives on ATs, specifically [15,16,26,28,32], but they do not provide a rich understanding of human perspective in this complex field. A qualitative approach is necessary to gain in-depth insights into passengers' opinions about ATs and enable the development of techniques to influence public trust and opinion. Therefore, this research aims to answer the following research question: *What factors influence passengers when evaluating the choice to ride a fully autonomous train, and what can the railway industry do to positively influence their perspectives?*

This paper's Sect. 2 summarizes the background and related work, Sects. 3 and 4 outline the study design and results, Sect. 5 discusses the main results, the limitations, and possible future work, and Sect. 6 concludes the paper.

2 Background and Related Work

Each transportation domain has standards to describe its levels of automation. For railways, these levels are known as Grades of Automation (GoAs) and have been defined in IEC 62290-1 by the minimum mandatory functions [21]. Figure 1 shows an overview of the GoAs and was also used to illustrate the concept to participants (see Sect. 3.2). This research focuses on railway transport at GoA 4; at this GoA, the train is fully autonomous and staff do not execute operational functions and may instead perform customer service tasks. Other than that, people are only involved in performing system monitoring from afar and maintenance services.

GRADE OF AUTOMATION	TRAIN OPERATION	SETTING TRAIN IN MOTION	DRIVING AND STOPPING	DOOR CLOSURE	OPERATION DURING DISRUPTION
1	ATP	Driver	Driver	Driver	Driver
2	ATP and ATO	Driver / Automatic	Automatic	Driver	Driver
3	DTO	Automatic	Automatic	Automatic / Attendant	Attendant
4	UTO	Automatic	Automatic	Automatic	Automatic

Fig. 1. Graphic differentiating between the GoAs for railways. Automatic Train Protection (ATP), Automatic Train Operation (ATO), Driverless Train Operation (DTO), Unattended Train Operation (UTO). Figure adapted from Alstom [3].

The terms automatic and autonomous are often considered synonymous, but the scope of autonomy is different. Automatic trains don't share tracks with anything but other automatic trains and are typically directly connected to the infrastructure, but autonomous trains may encounter driven or autonomous trains, cars, and pedestrians on their tracks. This requires autonomous trains to see ahead and around their track, communicate, and make decisions [3].

2.1 Perspectives on Autonomous Trains

Efforts have been made to understand people's perspectives on ATs. Fraszczyk, Brown, and Duan [15] discovered, via questionnaire, that people were mainly concerned about communication and potential technical features; fewer worries arose regarding the train design itself and system maintenance. Fraszczyk and Mulley [16] reported a slightly different picture in Sydney, AU, where participants reported concerns about passenger safety and rated a train driver's presence as important. While those results focus on concerns, a user acceptance survey of ATs in Germany showed that respondents were willing to use ATs in the future [32]. Free text entries indicated that participants who had ridden an AT felt more secure because the train could not deviate from the rails and the system was well secured. Still, some participants felt insecure because they did not fully trust the system and were uncertain about what to expect in case of disturbances, partly in line with [15,16].

Other research used questionnaires to create user acceptance models for public transport and train autonomy and investigated factors influencing use and acceptance. They highlighted constructs that push people both toward and away from acceptance, including habits, safety expectations, social-environmental factors such as social influence and facilitating conditions, and elements of personality and preference, like hedonic motivation and trust tendencies [26,28].

This paper builds on existing research by adding a comprehensive, qualitative overview of people's perceptions of ATs. We further investigated concerns and fears related to ATs as well as acceptance and potential solutions that prospective users foresee for the implementation of ATs. Our qualitative approach aims to better understand the reasons behind people's perceptions and to create a more extensive basis for a traveler-centered approach.

3 Study Design

To address the research question, two qualitative research methods, focus groups (FGs) and creative problem-solving (CPS) sessions, were used to yield rich, varied information from a passenger's perspective[1] [9,39]. FGs were selected due to the appeal of having direct conversations with people and the ability to explore multiple perspectives [19,29]. They aimed to discuss how passengers felt about ATs, identify concerns, and explore related topics. CPS was added due to its ability to generate solutions based on multiple individuals' perspectives and to complete an initial group evaluation of these ideas. This strategy is effective at generating many out-of-the-box ideas, can be a positive, engaging experience for participants, and can highlight the connections that participants make between different ideas related to the topic [45]. The CPS aimed to identify and evaluate what measures might make passengers feel safer and more comfortable riding ATs. The study was allowed to proceed by the Research Institute of Information and Computing Sciences based on an Ethics and Privacy Quick Scan.

3.1 Participants

Participants were recruited via a University Master-Level course and in-person contacts. They had to be older than 18, able to communicate and understand English on approximately a B1 level, and have ridden on a train. All participants received refreshments and snacks as compensation.

Four sessions, with a total of 26 participants (15 female, 9 male, 2 undisclosed), were held. Participants were between 22 and 32 years old (M = 24.6, SD = 2.3). Most were Masters students in Human Computer Interaction (17). Others were Masters students in Law (1), Literature (1), and an undisclosed study (1), Ph.D. candidates in the fields of AI (2), Psychology (1), and Human Computer Interaction (1), and a Resident Manager who specialized in retail. Two participants chose not to disclose their demographic information. See Table 1) for demographics of the sample.

[1] In session 1, the FG and CPS sessions were conducted on different days. As the methods complement each other, all other sessions combined the FG and CPS into one session with short breaks throughout to limit the recall required.

Table 1. Demographic summary of study participants. *Participants B, L, and G could not join the CPS session, and participant M could not join the FG.

Session	Size	Gender		Age		Undisclosed	Participants
		F	M	Mean	SD		
S1	6	3	3	23.0	1.1	0	A, B*, C, D, E, F
S2	6	4	0	24.0	1.9	2	G*, H, I, J, K, L*, M*
S3	6	3	3	23.7	0.8	0	N, O, P, Q, R, S
S4	7	4	3	27.0	2.4	0	T, U, V, W, X, Y, Z

3.2 Materials

The study was conducted in discussion-friendly environments on a university campus, and two researchers moderated each session. Figure 1 from Alstom was used to illustrate the GoAs and create a general understanding amongst participants. A two-minute video from a French rail logistics company was used to showcase a small selection of the capabilities of ATs [41].

Moreover, a script with the introduction, a list of six topics for the FG, two excursions for the CPS, and guidelines for the moderators was developed. Also, evaluation criteria were developed as a basis for assessing and choosing among participants' proposed ideas (see Table 2). These included three pre-selected criteria used in each CPS, Feasibility, Benefit, and Penalty of disregarding the feature, which were derived from Wiegers' Prioritization strategy [46]. Additionally, participants selected two additional criteria adapted from Dean et al. [11]. While the criteria evaluation was limited by time, the choices made with this constraint offer insights into participants' preferences and priorities[2].

Participants received sticky notes, pens, and optional name tags. Mobile phones were used to record the sessions. Transcription was supported by the Microsoft Word Web transcription tool, and coding was done in NVivo 20 [35].

3.3 Protocol

Each session took approximately 1 h and 45 min. First, moderators introduced themselves and the research. Participants read and signed a consent form regarding their participation and verbally consented to audio recording. The rest consisted of the focus group and the creative problem-solving session.

Part 1. First, definitions surrounding ATs were covered and Fig. 1, the image from Alstom explaining the grades of automation, was shown. After confirming clarity and allowing time for questions, six different topics, drawn from the literature on ATs, were prompted to spark discussion. These focused on (1) participants' initial impressions of ATs, (2) opinions on staff presence onboard, (3)

[2] Session 1 evaluated their ideas with different criteria, explained in Sect. 4.2.

Table 2. Idea evaluation criteria used at the end of CPS sessions.

	Criteria	Definition
Pre-selected	Feasibility	How possible is it to implement this idea in real life?
	Benefit	How high is the idea's benefit when implemented?
	Penalty	How high is the penalty of the idea not being implemented?
Optional	Effectiveness	To what extent will this idea achieve the objectives?
	Impact	How much will this idea impact passengers' experiences?
	Sustainability	How long will the impact last?
	Scalability	Is the idea able to be expanded and scaled?
	Stickiness	How much will the idea appeal to emotions?
	Legality	How legal would this idea be in a real-life setting?
	Cost	How much money, time, or other resources will it take to implement?

reactions to autonomy in various transportation modes, (4) the impact of ATs on cargo transport, (5) the influence of the AT capabilities video on their opinion, and (6) opinions on informing passengers about AT operation.

Part 2. Following a short break, the protocol for the CPS started with an introduction to the CPS methodology and the prompt *"What measures could train companies use to improve passenger trust and comfort in autonomous trains?"* Participants were instructed to brainstorm verbally as a moderator wrote their ideas on a visible whiteboard. After approximately five minutes, a role excursion was introduced. Participants were each assigned a role that interacted with an AT (e.g., passenger with a baby, pedestrian near the tracks, commuter) to pull people out of their paradigms by suggesting other priorities, meant to increase diversity of thought. Participants were asked to empathize with that role and share their thoughts. Afterward, the brainstorming resumed for another five minutes before the walking excursion wherein participants were prompted to take a five-minute walk around the building and record features of the building that made them feel safe. Once everyone returned and explained the measures they identified, the moderator urged another five minutes of brainstorming until a natural lull in idea generation. Afterward, the voting strategy was introduced. Moderators counted the number of ideas on the board and divided that number by two to determine the amount of "points". Each participant allocated their points to their preferred ideas without a limit on how many of those points could be given to a single idea, using sticky notes to limit peer influence on voting.

The top five ideas, based on points, were transferred to another area of the whiteboard for evaluation. The pre-selected criteria were explained, and a moderator walked the group through selecting two optional evaluation criteria from Table 2. Participants then scored each solution from one to five for each criterion to highlight those that may be most effective or beneficial. Lastly, if time allowed, the moderators checked if the participants wanted to assign weights to any of the criteria to emphasize their importance to the situation. The final score was calculated by adding the weighted score of each criterion. The session ended by thanking the participants and reminding them to contact the moderators with questions or concerns.

3.4 Analysis

The sessions were audio recorded, and the sticky notes and resulting whiteboards of the CPS session were photographed. The recordings were transcribed and anonymized to ensure statements and opinions could not be traced back to individual participants. The methods were first analyzed separately:

FG. The discussed topics were sorted into categories based on Grounded Theory and, more specifically, the principles of open and axial coding [7]. Codes were generated based on patterns found in the discussion. The researchers then developed thematic families and linked codes to get an overview of the main categories. The researchers coded the initial two sessions together, then defined the codes in a codebook to establish trend boundaries and ensure the coding was verifiable and replicable; the codebook can be requested from the researchers. A selection of codes, chosen based on how well they represent the conversations and the relative frequencies at which they were mentioned, were visualized.

CPS. The ideas on the whiteboards were transferred into a table and grouped based on their broader categories and, if applicable, subcategories, both developed based on the available data. Subsequently, a comparison was made between the categories, ideas, and session groups and summarized visually. Each group's top five ideas were connected to their origin in conversation and their discussions during the evaluation. The scores given to each idea cannot be compared to ideas outside of that group as they have been calculated using different selection criteria and weighed differently, meaning that the results do not have a baseline that would make this comparison meaningful. This also applies when looking at the average score of each session. Therefore, to draw comparisons between the sessions, the winning ideas themselves were compared to each other, enabling the identification of trends within and between each session.

Once this analysis was complete, the findings across all sessions and both methods were connected and correlated.

4 Results

This section presents the study results. Although both methods addressed the same research question, the information gathered had different focuses. The FG focused on identifying opinions and concerns regarding ATs, while the CPS focused on generating ideas that make passengers feel safer and more comfortable.

4.1 Focus Group

Participants covered topics beyond those prompted by the researchers including experiences with train systems, personal opinions, and hypotheses about their reactions to potential events. They examined concepts from multiple points of view, and some adjusted their perspectives as they talked with others. The discussions in the FGs are visualized with a combination spectrum display and a heat map, seen in Fig. 2. The depth of the discussion of each topic informs the heat map: the darkest colors were the most discussed topics. The categories in the visualization are summarized below.

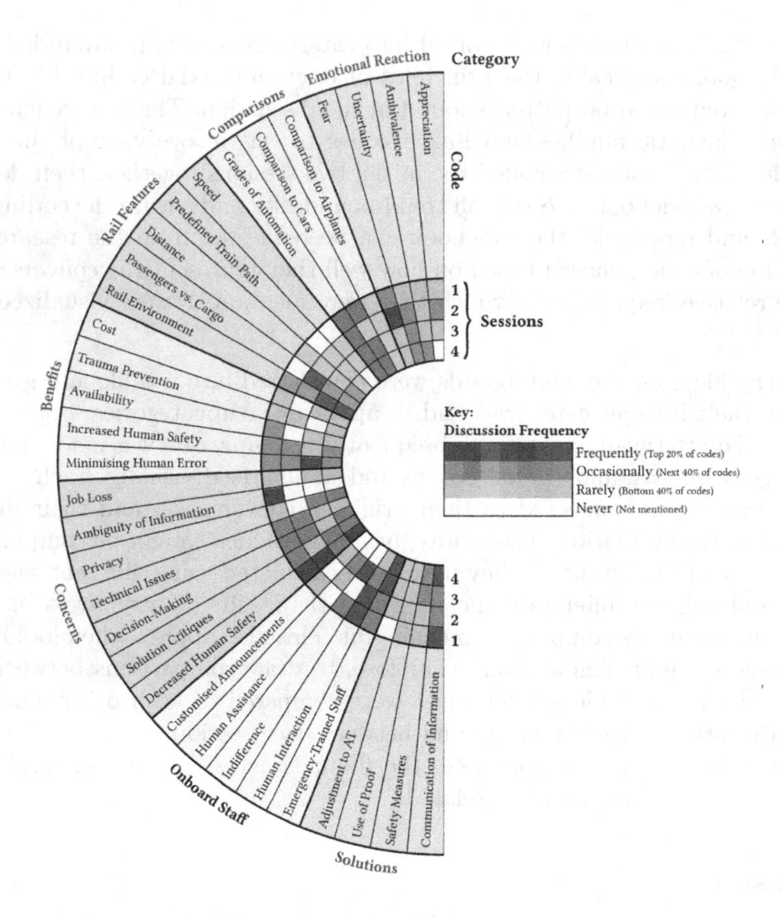

Fig. 2. Frequency of discussion related to certain topics for each FG session (1-4).

Emotional Reaction. Participants had diverse emotional reactions about several topics related to ATs. Nine expressed a neutral-positive reaction, stating that the autonomous nature of the trains would not deter them or that ATs were equivalent to standard trains concerning certain criteria. Three participants appreciated and trusted AT technology due to the minimized human error. Fear was another common emotion, generally in response to the lack of a human backup or monitoring presence. The most shared emotion was uncertainty, often related to topics such as Technical Issues, Decision-Making, and Privacy. Some participants, like Speaker X, clarified that they had similar levels of uncertainty with manual trains:

> *"I'm not familiar enough with what a train driver actually does to be able to gauge whether that's well automate-able and what could be potential dangers of that."*

Comparisons. Participants often used other GoAs and types of autonomous vehicles to discuss factors that impacted their perspective of ATs. Participants preferred that the train be accompanied by a human attendant in case of an emergency and compared attendants in ATs to pilots acting as a backup to the autopilot in airplanes. They preferred AT implementation over self-driving cars because the environment around the ATs is generally less complicated. Relatedly, participants considered trams to be the last priority for automation, as Speaker A stated, *"I would prefer automated metros and trains over trams."*

Rail Features. Discussion of this topic focused on the speed the trains traveled, the benefit of ATs having a set path, and the distance that an AT had to travel in total or between stations. Participants mentioned the rail environment, including the presence of non-autonomous trains and the possibility of cars, people, and animals on the tracks. ATs carrying cargo sparked concerns about increased weight and the consequent additional time required to stop. Speaker U mentioned unease about, *"what kind of stuff is in the train if things go south,"* a reference to the February 2023 train derailment in Ohio that spilled hazardous materials.

Benefits. Participants posited that ATs could have better safety outcomes for passengers, staff, and bystanders and could minimize the trauma drivers experience in collisions with people or animals by removing responsibility. The potential lower cost to both passengers and the rail system and the expectation that ATs would operate on a more consistent schedule were mentioned as factors that could encourage AT usage. A common topic across sessions was the potential for ATs to minimize human error by not being restrained by human capabilities, such as limited cognitive resources, decreases in sustained attention, poor decision-making, inefficient reaction times, and issues related to physical health (e.g., tiredness, sudden health problems). For example, Speaker J reasoned that:

> *"[ATs are] being [modeled] on human behavior, and on the best examples of human behavior. They're being trained based on that. So, technically, it should be better."*

Concerns. Some participants, generally those hesitant about ATs, debated the solutions suggested by other participants. The possible decrease in job opportunities was mentioned in three sessions. One session discussed the idea that success statistics and information about ATs could be manipulated by the train system for the false comfort of the passengers, to not *"scare away people,"* according to Speaker B. Participants had issues with the data monitoring solutions like cameras and behavioral analysis, questioning data collection, usage, and security. Participants in all sessions were also bothered by the possibility of bugs in the code, the risk of hacks, the integration of unmonitored human activity, and the impact of broken sensors on train operation. While participants agreed that the vision and reaction times of ATs were likely better than those of human drivers, they were also concerned about AT's decision-making capabilities in complicated or controversial situations. Speaker X contemplated the comprehensiveness of the AT's code: *"If something unexpected happens that wasn't, somehow, part of the training, then it might become dangerous."*. Speaker A shared similar fears:

> *"Computers are not the best at encountering unexpected situations and giving the best solutions for those types [of] scenarios."*

Participants in all sessions were worried about potential safety issues for passengers, staff, and bystanders, and believed that the public would perceive ATs as less safe than standard trains. Another worry was the possibility of passengers being left alone on an AT with malicious strangers. Speaker X elaborated:

> *"I would probably feel a bit uncomfortable simply because all types of humans enter the train, and there's also been cases in the news where people were assaulted or sexually harassed."*

Onboard Staff. The perceived presence of staff onboard was commonly a point of comparison between ATs and standard trains. Customized, non-standard announcements were an appreciated, memorable element of some participants' trips. Participants mentioned ways human staff assist passengers in ways an AT cannot. While some participants did not prioritize or notice onboard staff, the impact of staff-passenger interactions and the importance of a "human touch" were central in most sessions. Speaker B described that:

> *"It's about having a friendly voice or a friendly face, [...] sometimes it's someone that says hello, sometimes it's someone that you just don't even see, but you know they're there."*

For participants in all sessions, a clear priority was to have staff onboard trained to handle people in emergencies. They desired someone to manage passengers in emergency or conflict situations, administer basic first aid, and be an authority figure, as summarized by Speaker Z:

> *"[I would want] someone that can take over the machines, and that can take care of issues like the ones that [Speaker X] mentioned, like assaulting, problems with the people, and somebody that can take care of health issues, like say that somebody has a heart attack."*

Solutions. Participants made several suggestions during the FG to make ATs a more attractive option for passengers, including a period in which passengers could choose between standard trains and ATs. They also suggested that statistics and demonstrations could prove the efficacy, safety, and benefits of ATs to passengers. Participants also emphasized the importance of visible tools and strategies to increase passenger safety, including off-site monitoring, emergency phone numbers, and first aid equipment. Speaker P suggested: *"An emergency button on the train that you can press, and having the device that you use for first aid, the [defibrillator]."* Another popular topic was how much and what types of information should be communicated to passengers, including the need for clear safety protocols and materials that enable passengers to make informed decisions. Participants stressed that train companies must be thoughtful about the information and statistics shared to balance transparency and the possibility of scaring anxious passengers. Participants, including Speaker W, clarified that it was also important for *"the people around, [in] the environment, to be aware that the train is an autonomous train."*

4.2 Creative Problem Solving

During the CPS sessions, participants generated a total of 165 different ideas: 33 in Session 1, 40 in Session 2, 48 in Session 3, and 44 in Session 4. Ideas were grouped into 25 named and numbered subcategories and further consolidated into seven titled categories. Subcategories with the label *No subcategory* indicate that the ideas within are linked directly to the superordinate category as they could not be otherwise merged. Figure 3 shows the number of ideas generated in each session by category. Table 3 outlines the categories, their subcategories, and the number of generated ideas within each subcategory across all sessions.

Table 3 reveals that "Train Experience", "Safe Passenger Transport", "Changing Opinions", and "Information" encompass the majority of ideas. Several of these categories are represented at least once in the top ideas of each session. These subcategories capture the most ideas: *14. Include additional train*

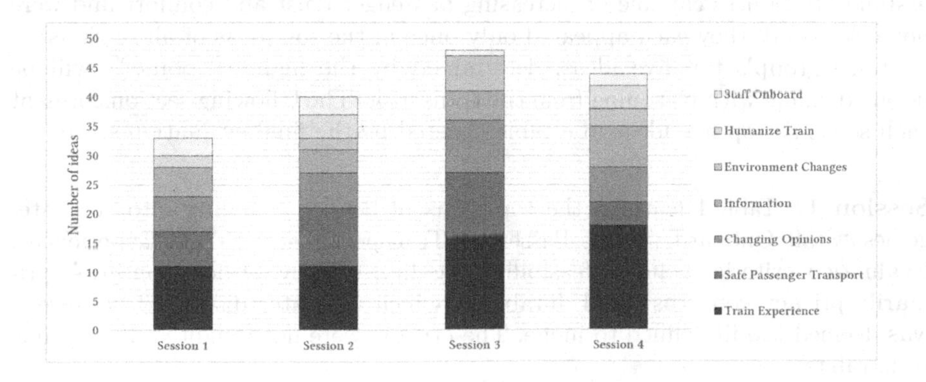

Fig. 3. The number of ideas generated in each session by category.

Table 3. Categories and subcategories of the CPS ideas.

Category	Subcategory		# of ideas
Safe Passenger Transport	1	Display safety information	10
	2	Monitor passengers	5
	3	Implement safety features	5
	4	Offer protection against other passengers	4
	5	Enable emergency contact	4
	6	Make the train seem more reliable	3
Information	7	Present opportunities to learn more	9
	8	Update passengers with train status	4
	9	Indicate train is autonomous	3
	10	Provide next stop announcements	3
	11	Give updates about the environment	2
Staff Onboard	12	*No subcategory*	7
	13	Employ dogs	2
Train Experience	14	Include additional train features	20
	15	Improve the ambiance	12
	16	Include decorations or themed train cars	12
	17	Offer food	7
	18	Provide comfortable seats	5
Changing Opinions	19	Advertise the trains	14
	20	Use proof	8
	21	Award giveaways	5
Humanize Train	22	Humanize autonomous conductor	8
	23	Implement autonomous conductor on app	2
Environment Changes	24	*No subcategory*	7
	25	Improve transportation apps	4

features, *19. Advertise the trains*, *15. Improve the ambiance*, *16. Include decorations or themed train cars*, and *1. Display safety information.* Despite the many ideas related to these subcategories, during the voting and evaluation, they were assumed to be less capable of increasing passenger trust and comfort and were not prioritized; they each appeared only once in the top ideas of all sessions.

Each group's top five ideas, determined by the highest "points", will be detailed, along with reasoning from the transcript. The following sections present each session's top five ideas in a table ordered by the final evaluation score.

Session 1. Table 4 features the top ideas of Session 1 falling into the categories "Staff Onboard", "Safe Passenger Transport", and "Train Experience". Evaluation criteria included feasibility, legality, efficacy, ethical concerns (primarily privacy concerns), and durability, which was later discarded because it was deemed too ill-defined to judge. The criteria were not weighted due to time constraints.

Idea: "Visible cameras only in main area"|. Visible cameras seemed to make participants feel safer. They considered ethical and privacy issues that

Table 4. Session 1's top five ideas, including category name, subcategory number, and total rating received during the evaluation.

Category	Subcat.	Idea name	Total
Safe Passenger Transport	2	Visible cameras only in main area	17.5
Staff Onboard	12	Having responsible person on train	16.5
Train Experience	15, 18	Well-lit, comfortable seats	15.0
Safe Passenger Transport	3	Manual over-ride (emergency brake)	15.0
Safe Passenger Transport	6	Reduce other malfunctions	14.0

total surveillance of the entire train interior could bring, how recordings would be stored, and who would have access. A participant contended that cameras should be in the main areas, not everywhere, but others argued that passengers might feel safer if the cameras covered everything. According to Speaker A, gaps in surveillance could be dangerous: *"Someone who really wants to do something, and really pays attention, could find a blind spot and do something."*

Idea: "Having responsible person on train"|. Participants stated that having identifiable, visible, and responsible personnel on the AT to take charge if necessary would be reassuring. They posited that there would be no extra costs since employees are currently deployed on each train. However, some were concerned about the people who would be solely responsible in risky situations.

Idea: "Well-lit, comfortable seats"|. This suggestion came from a participant who valued a comfortable journey over efficiency. They posited that the experience of riding an AT offers more uncertainties and that having comfort in their immediate surrounding is thus more important.

Idea: "Manual override (emergency brake)"|. When discussing the benefits of a panic button, participants developed the idea of having a manual override or emergency brake passengers could use in case of emergency. During the evaluation, the risk of passenger misuse was addressed with the caveat that the feature would not give a passenger control beyond an emergency stop.

Idea: "Reduce other malfunctions"—. Participants stated technical malfunctions on an AT would be scary. They suggested a backup system and explicit communication from the AT to the passengers that it is aware of the issue and the plans for resolution. Speaker E explained:

> *"If I were on an autonomous train and something like the lights, or the Internet, or whatever, malfunctions, I think I would be way more worried than if I'm on a normal train and the lights flicker."*

Session 2. Table 5 features the top five ideas of Session 2 belonging to the categories "Humanize Train", "Information", "Safe Passenger Transport", "Changing Opinions", and "Train Experience". Session 2 evaluated their ideas according to the predefined criteria in Table 2. Additionally, stickiness was considered,

Table 5. Session 2's top five ideas, including category name, subcategory number, and total rating received during the evaluation.

Category	Subcat.	Idea name	Total
Safe Passenger Transport	5	Contact someone outside of AT	21.00
Information	7	Information about operation of AT	20.75
Humanize Train	22	Humanize autonomous conductor	16.00
Train Experience	14, 16	Futuristic train setting	16.00
Changing Opinions	20	Use statistics	15.75

because the goal of establishing trust targets passengers' emotions, and scalability, allowing participants to consider implementation outside the Netherlands. While weighing the criteria, the penalty criteria was emphasized by 1.5 and scalability was halved because they aimed to stress applicability to the Netherlands.

Idea: "Contact someone outside of AT"|. The ability to contact somebody outside of the vehicle (e.g., through video chat screens mounted on train seats) in case of an emergency was proposed to give passengers more resources. In the evaluation, participants considered it feasible but emphasized the possibly detrimental necessity of an adequate phone signal.

Idea: "Information about operation of AT"|. Participants felt passengers should be informed about AT operation and emergency procedures but advised caution when sharing information to avoid confusion or anxiety.

Idea: "Humanize autonomous conductor"|. Participants suggested personifying the AT by giving the autonomous conductor a name and personality. They added that the train's character could change with each train or per person, as passengers might have different interaction preferences. During the evaluation, participants pointed out that the perception of this idea is strongly subjective, and its efficacy may differ depending on the passengers' culture. Speaker K added that the impact could be big as *"it's, like, a pure appeal to emotion."*

Idea: "Futuristic train setting"|. To increase passengers' comfort and enjoyment in an AT, a participant suggested making the AT futuristic by having charging points, videos, and somewhere passengers could lie down. Others expressed that this idea might be very expensive and require new trains to be built, which would take additional time and could delay technological advances.

Idea: "Use statistics"|. Participants suggested using statistics to portray the train as a safer transit option. They suggested comparing the fatality rate of passengers on ATs to that of other modes of transportation. Additionally, Speaker K explained that if *"those trains were running on time more than the other ones, then people would think it's a big improvement."*

Session 3. Table 6 contains the top ideas from Session 3, which fall into the categories of "Information", "Staff Onboard", and "Safe Passenger Transport".

Table 6. Session 3's top five ideas, including category name, subcategory number, and total rating received during the evaluation.

Category	Subcat.	Idea name	Total
Safe Passenger Transport	3	Emergency button	25.00
Staff Onboard	12	Having an attendant onboard	24.00
Information	7	Informative videos how the train works	20.00
Information	8	Train status (speed, map, ETA)	20.00
Safe Passenger Transport	4	"Shhh" button	17.00

Session 3 evaluated their ideas according to the predefined criteria in Table 2 as well as effectiveness and cost, seen as a way to ensure passenger comfort while simultaneously minimizing expense. They decided to give twice as much weight to effectiveness over the other criteria.

Idea: "Emergency button"|. The idea of one emergency button evolved into multiple buttons that indicate the type of emergency. Participants had concerns and suggested the buttons would need to be subtle to minimize costly misuse for emergencies involving first responders. They suggested implementation might be complicated as the buttons should connect to external resources.

Idea: "Having an attendant onboard"|. An attendant onboard would handle emergencies and serve as a link between the AT and passengers. Speaker Q recalled an incident when their arm was stuck between train doors; having an attendant would reassure passengers that the AT would not move if passengers were in danger. However, concerns were raised about the high cost of attendants due to training and the counter-intuitive nature of adding staff, since avoiding paying and managing staff was a key reason for automation.

Idea: "Informative videos of how the train works"|. A participant suggested providing passengers with informative videos about how ATs work. During the evaluation, a concern was raised that the level of information could be too high or the videos could be uninteresting, leading to no benefit. Nevertheless, they saw it as an important feature when prioritizing transparency.

Idea: "Train status (speed, map, ETA)"|. To improve passengers' trust, a participant suggested using existing screens to show live feedback on the train's status. The screen could cycle through information like the next station, arrival time, speed, and temperature. During the evaluation, participants said the information display should be updated in real-time, indicate delays, and assist passengers in getting onto the correct trains throughout their journey.

Idea: "Shhh button"|. One participant suggested a button that voices "shhh" to indicate to quiet rowdy passengers, primarily to help those with anxiety or fear of retaliation. During the evaluation, others argued that passengers could approach others directly, and that misuse could be high.

Table 7. Session 4's top five ideas, including category name, subcategory number, and total rating received during the evaluation.

Category	Subcat.	Idea name	Total
Safe Passenger Transport	1	Clear safety exits and safety tools	29.00
Staff Onboard	12	Human able to manage train, people, health	24.50
Staff Onboard	12	Visible, touring responsible person	21.75
Changing Opinions	19	Reduced price	15.00
Train Experience	17	Food services accessible without electricity	14.25

Session 4. Table 7 features the top ideas of Session 4 belonging to the categories "Staff Onboard", "Safe Passenger Transport", "Changing Opinions", and "Train Experience". Session 4 evaluated their ideas according to the predefined criteria in Table 2 as well as effectiveness, focusing on the idea's impact to determine an idea's worth, and cost, a seemingly necessary and realistic dimension. The weight of the ratings for feasibility and penalty were increased by 1.5.

Idea: "Clear safety exits and safety tools"|. Participants clarified that, even if they are not needed, having indications of safety instructions and tools like fire extinguishers and smoke detectors would reassure passengers. Speaker X argued that having this is important because people need to know:

> *"what to do in case of emergency, where fire extinguishers are, maybe even more than on a train where there is personnel because they would know. But if there's no personnel, it should be super clear to the passengers where the emergency exits are, or fire[-fighting] stuff."*

Idea: "Human able to manage train, people, health"|. This idea originated with Speaker Z, who said that a staff member able to take control of the machine, take care of human behavior problems, and manage health issues was a priority. Later on, however, Speaker T argued against the feasibility of this idea, suggesting that there may be a limit on the pool of qualified personnel.

Idea: "Visible, touring responsible person"|. This idea extends the idea above, emphasizing the visibility of the train staff and clarifying that they would walk around to reassure passengers. This idea was critiqued as performative and others argued it is not important to see them as long as they do their job.

Idea: "Reduced price"|. Participants suggested that a notably reduced price could encourage people to try out the ATs and that the experience would increase trust in ATs. However, participants doubted savings would be passed to the passengers instead of considered profits, and others worried low fares could negatively affect the train companies' longevity and reliability. The penalty of not implementing reduced prices was deemed low because people would likely take the train either way. Speaker T was hesitant about the idea's feasibility:

"Implementing all those things, and building new trains, maybe also adapting the railway systems... It would cost so much that they would take a lot of time to get the money they spend back into a return on investment."

Idea: "Food services accessible without electricity"|. In response to a participant's experience with a train that stopped without refreshment access, they suggested services that operate when the train's electricity fails. However, they noted limited value due to the infrequency of electrical system failures.

Winning Ideas Across Groups. Within each session's top ideas, the winning idea was in the category "Safe Passenger Transport"; two sessions had more than one highly-voted idea in this category. All groups except Session 2 had an idea in the category "Staff Onboard"; each rated as second-highest in those three groups. Additionally, all groups except Session 4 had an idea in the category "Information", and all groups except Session 3 had an idea in the category "Train Experience". Further trends emerge in the subcategories. Ideas within subcategory *3. Implement safety features*, was developed by Sessions 1 and 3, ideas from subcategory *7. Present opportunities to learn more* was chosen by Sessions 2 and 3, ideas from *8. Update passengers with train status* was chosen by Sessions 1 and 3, and ideas from *19. Advertise the trains* was chosen by Sessions 3 and 4.

5 Discussion

From the emphasis on interactions with other elements on the tracks of the rail environment and concerns about decision-making, a pattern of distrust in ATs to handle complex environments was evident. Simultaneously, there was a shared belief that ATs could minimize human error through sensors and constant monitoring of the environment, a presumption supported by research [14,40]. Participants' prioritization of safety over comfort is apparent in the results from the CPS sessions; the overall winning ideas of all sessions were ideas from the category "Safe Passenger Transport". This highlights the importance of addressing basic travelers' needs first, specifically'reliability' and'safety,' as suggested previously by van Hagen and van Oort [18]. It also indicates that participants hesitate to trust an AT and its capabilities without adequate knowledge and transparency from the train system about its operation and how the train will protect and help passengers in need.

Another important topic that all groups featured in their top-ranked ideas was the need for human staff onboard, supported by Fraszczyk et al. [15]. Every session suggested some variation of the idea, and three out of the four groups prioritized "Staff Onboard" as the second-highest idea. This was also reflected in the hesitation expressed in the FGs by several participants to the idea of having no responsible figure onboard who could intervene and take control in emergencies. Passengers should not feel helpless in an emergency and should be able to contact and receive help from somebody equipped to handle situations inside the train or with external resources.

Discussions about uncertainty were often intertwined with the topics of Technical Issues, Decision-Making, and Privacy. This indicates that these are key areas requiring further explanation when implementing ATs, as a lack of information seems to increase anxiety. This is also reflected in the many ideas within the "Information" category and the fact that three of the four groups rated ideas from that category of top importance. Further, two ideas from subcategory *7. Present opportunities to learn more* were included in the final ranking. Together, this indicates that train systems should provide clear information about the nature of ATs so people can make informed decisions and have the option to access detailed information about AT operation and safety measures.

A final facet participants theorized could improve passenger trust and comfort is to address their hedonic response to, or physical experience of, the train environment. Nearly a third of the ideas fell into the category "Train Experience", and three of the four groups voted ideas from this category into their top five. Whether this was related to accessibility, entertainment, more comfortable seats, food availability, decorations, or improved cleanliness, participants believed that the sensory experience of the train ride would impact passengers' trust and comfort in ATs. This was seemingly due to the assumption that a higher-quality experience is associated with higher-quality technology. Therefore, prioritizing this component will likely improve the traveler experience. Yet it should be noted that opinions about how to develop the best experience will vary based on culture and individual [26,28] and that enhancing the experience likely becomes more relevant after addressing the basic needs of travelers, as illustrated by van Hagen et al. [18].

Conclusively, the discussions in this study support earlier findings from the literature. Our participants were more concerned about their safety when in an AT compared to a train being driven by a human, and questioned the AT reliability and what would happen if the train encountered problems it would not be equipped to handle. Other topics that were mentioned in our study are also reflected in the literature, namely the possibility of job loss, concerns about the reliability of the sensors and software that runs ATs, and worries about how secure the AT system is against malicious attacks. The generated ideas, especially the highest-ranked ones, mostly focus on overcoming or minimizing the probability of these concerns becoming reality.

This study generated ideas that, when implemented, could help alleviate these concerns and therefore make ATs more accepted by the public. The relative consistency of our data across sessions implies that these concerns are not only individual positions but also indicative of a larger population of passengers and bystanders. Therefore, while it may not be generalizable on its own, our study contributes to the growing research regarding ATs, especially with regard to the focus on passenger concerns.

5.1 Limitations

Several limitations impact this study. One potential bias might arise from the participants' relative similarity in age, education level, and study area. However, they were likely potential passengers of ATs, their opinions were strongly varied, and they may be more critical of technology than the average potential passenger.

Another potential source of bias is the prompts used during the study. One of the CPS results, a hesitancy to trust ATs due to the frequency of ideas in the category "Safe Passenger Transport", may be a result of the moderators' encouragement to think of measures that could improve passenger trust and safety. However, this topic also appeared frequently during the FG sessions without being prompted, indicating that this aspect is still important to participants.

When evaluating and contextualizing the study's results, it is important to note that participants mainly discussed hypothetical situations. Their beliefs about a hypothetical future may differ from a lived experience with automation, which could change passengers' trust, perceived control, and use [17]. Nonetheless, these discussions align with related literature and are still representative of possible passengers' thought processes about their trust in ATs.

Another possible limitation is the difference between Session 1 and the other sessions. Session 1 featured less well-defined and consistent evaluation criteria compared to the other sessions and the voting was conducted directly on the whiteboard, which made the numbers hard to read and might have inserted peer influence. This resulted in a mistake when identifying the top five ideas: "Well-lit, comfortable seats", with eight points, was picked instead of "Detailed info about speed, location, system status", which had nine points. Nevertheless, the opinions and ideas gathered from Session 1 are similar to those from other sessions. Other factors limiting our research included participants who had to leave during Sessions 1 and 2. This might have led to losing valuable information, opinions, and ideas. However, we still received valuable input from these participants and never dropped below five participants in a session.

5.2 Future Work

As one limitation of this study was the relatively homogeneous participant sample, future research could include more diverse and representative groups of people, with participants that differ in age, occupation, and expertise. This could yield new opinions and ideas based on different knowledge levels, life experiences, and needs, enriching these findings. An additional focus could be exploring the impact of existing knowledge and experience with ATs. By running a similar study with groups of people who are educated about ATs, it would be possible to identify the concerns that disappear with experience and those that remain.

Another strategy could include experts who evaluate the feasibility of user-provided solutions developed during the CPS. They could assess proposed solutions, identify hidden problems for implementation, appraise the usability, and identify new user concerns or feedback. Furthermore, winning ideas from the CPS could be trialed and their effect on perceptions of safety and comfort could be measured. These approaches would show if the ideas that were generated by the participants could be implemented in a real-life setting and if they would provide the desired impact.

6 Conclusion

For this paper, we conducted focus groups and creative problem-solving sessions to answer the research question, *"What factors influence passengers when evaluating the choice to ride a fully autonomous train, and what can the railway industry do to positively influence their perspectives?"*. These factors include ensuring safety for passengers and bystanders, the presence of emergency-trained staff, and the communication of information about safety and the technology within autonomous trains. This should address common questions, worries, and safety matters, empowering passengers to make informed decisions about their ridership. These concepts were mentioned multiple times during each focus group, highlighting the impact these factors have on potential passengers when evaluating the decision to ride an autonomous train. Furthermore, during the creative problem-solving sessions, the generated ideas that were voted into the evaluation stage primarily related to these factors, and the winning ideas especially prioritized safety requirements. The results of this study showcase the participants' need for consideration and assurance in all of the above-mentioned factors and contribute to the growing body of research in the field of user perspectives on autonomous vehicles, namely in the field of autonomous railway vehicles. Innovation does not follow a track, but are excited for future development in this research field.

References

1. Alessandrini, A., Alfonsi, R., Delle Site, P., Stam, D.: Users' preferences towards automated road public transport: results from European surveys. Transport. Res. Procedia **3**, 139–144 (2014)
2. Alstom: Alstom demonstrates fully autonomous driving of a shunting locomotive in the netherlands (2022). https://www.alstom.com/press-releases-news/2022/11/alstom-demonstrates-fully-autonomous-driving-shunting-locomotive-netherlands
3. Alstom: Autonomous mobility: The future of rail is automated (2022). https://www.alstom.com/autonomous-mobility-future-rail-automated
4. Azad, M., Hoseinzadeh, N., Brakewood, C., Cherry, C.R., Han, L.D.: Fully autonomous buses: a literature review and future research directions. J. Adv. Transport. 1–16 (2019)
5. Bahn, D.: Weltpremiere: Db und siemens präsentieren ersten automatisch fahrenden zug (2021). https://www.deutschebahn.com/de/presse/pressestart_zentrales_uebersicht/Weltpremiere-DB-und-Siemens-praesentieren-ersten-automatisch-fahrenden-Zug--6867098
6. Bansal, P., Kockelman, K.M., Singh, A.: Assessing public opinions of and interest in new vehicle technologies: an austin perspective. Transport. Res. Part C: Emerg. Technol. **67**, 1–14 (2016)

7. Blandford, A., Furniss, D., Makri, S.: Qualitative hci research: Going behind the scenes. Synth. Lect. Human-Center. Inf. **9**(1), 1–115 (2016)
8. BMW: Autonomous driving - five steps to the self-driving car (2020). https://www.bmw.com/en/automotive-life/autonomous-driving.html
9. Brewer, J., Hunter, A.: Foundations of Multimethod Research: Synthesizing Styles. Sage, Thousands Oaks (2006)
10. Craig, J., Nojoumian, M.: Should self-driving cars mimic human driving behaviors? In: Kromker, H. (ed.) HCII 2021. LNCS, vol. 12791, pp. 213–225. Springer, Heidelberg (2021). https://doi.org/10.1007/978-3-030-78358-7_14
11. Dean, D.L., Hender, J., Rodgers, T., Santanen, E.: Identifying good ideas: constructs and scales for idea evaluation. J. Assoc. Inf. Syst. **7**(10), 646–699 (2006)
12. Distler, V., Lallemand, C., Bellet, T.: Acceptability and acceptance of autonomous mobility on demand: the impact of an immersive experience. In: Proceedings of the 2018 CHI Conference on Human Factors in Computing Systems, pp. 1–10 (2018)
13. Dong, X., DiScenna, M., Guerra, E.: Transit user perceptions of driverless buses. Transportation **46**, 35–50 (2019)
14. Folsom, T.C.: Social ramifications of autonomous urban land vehicles. In: 2011 IEEE International Symposium on Technology and Society (ISTAS), pp. 1–6 (2011). https://doi.org/10.1109/ISTAS.2011.7160596
15. Fraszczyk, A., Brown, P., Duan, S.: Public perception of driverless trains. Urban Rail Transit **1**, 78–86 (2015)
16. Fraszczyk, A., Mulley, C.: Public perception of and attitude to driverless train: a case study of Sydney, Australia. Urban Rail Transit **3**(2), 100–111 (2017)
17. Gold, C., Körber, M., Hohenberger, C., Lechner, D., Bengler, K.: Trust in automation-before and after the experience of take-over scenarios in a highly automated vehicle. Procedia Manuf. **3**, 3025–3032 (2015)
18. van Hagen, M., van Oort, N.: Improving railway passengers experience: two perspectives. J. Traffic Transport. Eng. **7**(3), 2142–2328 (2019)
19. Hennink, M., Hutter, I., Bailey, A.: Qualitative Research Methods. Sage, Thousands Oaks (2020)
20. Howard, D., Dai, D.: Public perceptions of self-driving cars: the case of Berkeley, California. In: Transportation Research Board 93rd Annual Meeting, vol. 14, pp. 1–16. The National Academies of Sciences, Engineering, and Medicine Washington, DC (2014)
21. IEC: Railway applications – urban guided transport management and command/control systems – part 1: System principles and fundamental concepts. Standard IEC 62290-1:2014, International Electrotechnical Commission, Geneva, CH (2014). https://webstore.iec.ch/publication/6777
22. International, M.R.: Driverless trains introduced on Paris metro line 4 (2022). https://www.railwaygazette.com/metros/driverless-trains-introduced-on-paris-metro-line-4/62546.article
23. International, M.R.: Lyon metro line b goes driverless with goa4 upgrade (2022). https://www.railwaygazette.com/metros/lyon-metro-line-b-goes-driverless-with-goa4-upgrade/61963.article
24. International, R.G.: Automated main line train operation requirements to be studied (2020). https://www.railwaygazette.com/research-training-and-skills/automated-main-line-train-operation-requirements-to-be-studied/57870.article
25. Kaur, K., Rampersad, G.: Trust in driverless cars: investigating key factors influencing the adoption of driverless cars. J. Eng. Tech. Manag. **48**, 87–96 (2018)

26. Korkmaz, H., Fidanoglu, A., Ozcelik, S., Okumus, A.: User acceptance of autonomous public transport systems: extended utaut2 model. J. Public Transp. **24**, 100013 (2022)
27. Kyriakidis, M., Happee, R., de Winter, J.C.: Public opinion on automated driving: results of an international questionnaire among 5000 respondents. Transport. Res. F: Traffic Psychol. Behav. **32**, 127–140 (2015)
28. Madigan, R., Louw, T., Wilbrink, M., Schieben, A., Merat, N.: What influences the decision to use automated public transport? using utaut to understand public acceptance of automated road transport systems. Transport. Res. F: Traffic Psychol. Behav. **50**, 55–64 (2017)
29. Morgan, D.L., Krueger, R.A., King, J.A.: The Focus Group Guidebook. Sage, Thousands Oaks (1998)
30. Nordhoff, S., de Winter, J., Madigan, R., Merat, N., van Arem, B., Happee, R.: User acceptance of automated shuttles in berlin-schöneberg: a questionnaire study. Transport. Res. F: Traffic Psychol. Behav. **58**, 843–854 (2018)
31. Othman, K.: Public acceptance and perception of autonomous vehicles: a comprehensive review. AI Ethics **1**(3), 355–387 (2021)
32. Pakusch, C., Bossauer, P.: User acceptance of fully autonomous public transport. In: ICE-B, pp. 52–60 (2017)
33. Payre, W., Cestac, J., Delhomme, P.: Intention to use a fully automated car: attitudes and a priori acceptability. Transport. Res. F: Traffic Psychol. Behav. **27**, 252–263 (2014)
34. Piao, J., McDonald, M., Hounsell, N., Graindorge, M., Graindorge, T., Malhene, N.: Public views towards implementation of automated vehicles in urban areas. Transport. Res. Procedia **14**, 2168–2177 (2016)
35. QSR International Pty.: Nvivo 20 (2022). https://www.qsrinternational.com/nvivo-qualitative-data-analysis-software/home
36. Salonen, A.O.: Passenger's subjective traffic safety, in-vehicle security and emergency management in the driverless shuttle bus in finland. Transp. Policy **61**, 106–110 (2018)
37. Schoettle, B., Sivak, M.: A survey of public opinion about autonomous and self-driving vehicles in the us, the UK, and Australia. Technical report, University of Michigan, Ann Arbor, Transportation Research Institute (2014)
38. Parida, S., Franz, M., Abanteriba, S., Mallavarapu, S.: Autonomous driving cars: future prospects, obstacles, user acceptance and public opinion. In: Stanton, N. (ed.) AHFE 2018. LNCS, vol. 786, pp. 318–328. Springer, Heidelberg (2019). https://doi.org/10.1007/978-3-319-93885-1_29
39. Silverman, D., Marvasti, A.: Doing Qualitative Research: A Comprehensive Guide. Sage, Thousands Oaks (2008)
40. Singh, P., Dulebenets, M.A., Pasha, J., Gonzalez, E.D.R.S., Lau, Y.Y., Kampmann, R.: Deployment of autonomous trains in rail transportation: current trends and existing challenges. IEEE Access **9**, 91427–91461 (2021). https://doi.org/10.1109/ACCESS.2021.3091550
41. SNCF, G.: Autonomous train, the revolution on rails. Video (2022). https://www.youtube.com/watch?v=g-w7QuIYoG8
42. voor de Statistiek, C.B.: Hoeveel wordt er met het openbaar vervoer gereisd? (2023). https://www.cbs.nl/nl-nl/visualisaties/verkeer-en-vervoer/personen/openbaar-vervoer
43. Synopsys: The 6 levels of vehicle autonomy explained—synopsys automotive (2022). https://www.synopsys.com/automotive/autonomous-driving-levels.html

44. Times, N.: Ns running fewer trains due to staff shortages; disabled travelers heavily impacted (2022). https://nltimes.nl/2022/11/07/ns-running-fewer-trains-due-staff-shortages-disabled-travelers-heavily-impacted
45. Treffinger, D.J., Isaksen, S.G., Stead-Dorval, K.B.: Creative Problem Solving: An Introduction. Routledge, Abingdon (2023)
46. Wiegers, K.: First things first: prioritizing requirements. Softw. Dev. **7**(9), 48–53 (1999)
47. Wintersberger, P., Frison, A.K., Riener, A.: Man vs. machine: comparing a fully automated bus shuttle with a manually driven group taxi in a field study. In: Adjunct Proceedings of the 10th International Conference on Automotive User Interfaces and Interactive Vehicular Applications, pp. 215–220 (2018)
48. Zasiadko, M.: Fully automated metros run in six eu countries — railtech.com (2019). https://www.railtech.com/infrastructure/2019/11/19/fully-automated-metros-run-in-six-eu-countries/?gdpr=deny&gdpr=deny

Beyond Acceptance Models: The Role of Social Perceptions in Autonomous Public Transportation Acceptance

Nina Hieber[1](✉), Diana Fischer-Pressler[1,2], Monika Pröbster[4], Janika Kutz[1,3], and Nicola Marsden[4]

[1] Fraunhofer IAO, Nobelstraße 12, 70569 Stuttgart, Germany
{nina.hieber,diana.fischer-pressler}@iao.fraunhofer.de
[2] Frankfurt University of Applied Sciences, Nibelungenplatz 1, 60318 Frankfurt Am Main, Germany
[3] University of Kaiserslautern- Landau, Erwin-Schrödinger-Straße 57, 67663 Kaiserslautern, Germany
[4] Hochschule Heilbronn, Max-Planck-Street 39, 74081 Heilbronn, Germany

Abstract. The future of public transportation is on the verge of a transformative leap due to advances in automation and artificial intelligence. However, for (semi-)autonomous electric buses (AEBs) to reach their full potential in public transport, widespread public adoption is crucial. While acceptance models like UTAUT and TAM emphasize rational factors, our study uncovers the overlooked societal perceptions shaping AEB acceptance. Leveraging the Stereotype Content Model (SCM), we reveal nuanced attitudes towards AEB users and non-users. Our findings suggest that societal perceptions, often driven by automatic processes, significantly influence acceptance beyond rational considerations. Bridging this gap between technological innovation and social acceptability is pivotal for fostering inclusive, sustainable urban transport infrastructures.

Keywords: User Acceptance · Social Acceptability · Stereotype Content Model · Autonomous Shuttle · Public Transport

1 Introduction

Advancements in automation and artificial intelligence have the potential to revolutionize public transportation. Urban centers are experimenting with the integration of autonomous electric buses into their public transport ecosystems (e.g., [1, 2]). (Semi-) Autonomous electric buses, henceforth referred to as AEBs are vehicles equipped with hard- and software to enable autonomous driving, albeit with provisions for human intervention and remote control. The primary advantage of autonomous driving lies in the vehicle's ability to anticipate events, predict and respond to dynamic traffic conditions, and communicate with both urban infrastructure and other vehicles. These features position AEBs as crucial components in the development of smart cities. Cities that successfully mitigate traffic congestion linked to public transportation are set to experience significant enhancements in their quality of life and a reduction in energy consumption [3].

H. Krömker (Ed.): HCII 2024, LNCS 14733, pp. 26–39, 2024.
https://doi.org/10.1007/978-3-031-60480-5_2

The potential of AEBs in public transport can only be realized if they are widely adopted by the public as a means of transport. Public acceptance of AEBs has attracted scholarly attention, and research on AEBs has shown that public acceptance depends on several factors, such as ease of use or perceived usefulness (e.g., [4, 5]). However, existing research has primarily focused on technology acceptance without fully exploring the underlying societal perceptions that shape public attitudes towards autonomous vehicles [6, 7]. In response to this gap in the literature, this paper seeks to take a closer look at the psychosocial aspects of AEB acceptance, using the stereotype content model (SCM) [8, 9] as a framework for understanding social acceptability, offering insights into the mechanisms underlying perceptions in social spheres. Our research aims to shed light on the intuitive dimensions of AEB acceptance, thereby laying the groundwork for more effective strategies to promote their adoption in public transportation systems. By bridging the gap between technological innovation and societal acceptance, we hope to contribute to the realization of a more efficient, sustainable, and inclusive urban transport infrastructure.

2 Related Work

To understand people's perception regarding an AEB, this section presents related research and the foundational concepts guiding our study. We first shortly introduce prior research on AEBs acceptance, then elaborate on the aspects of social acceptability and on the SCM to examine social perceptions.

2.1 The Technology Acceptance of AEBs

People's intention to use an AEB is widely studied through the lens of acceptance models such as the technology acceptance model (TAM; e.g., [4]), the unified theory of acceptance and use of technology (UTAUT, e.g., [10]) and its extended version (UTAUT2, e.g., [1]). Such individual-level technology acceptance models incorporate factors regarding the evaluation of the technology, e.g., ease of use, performance expectancy, or hedonic motivation, as predictors of people's intention to use a new technology, which is in our study traveling with an AEB [11]. However, these models focus mostly on rational reasons for the acceptance of technology. While research on the acceptance of AEBs recognizes the importance of social factors, such as the influence of the opinions of significant others, on an individual's decision to use an AEB [1, 2], it overlooks the broader societal perceptions surrounding technology adoption. However, societal perceptions influence social acceptance, which has been identified as a key factor in system acceptance [12]. This oversight concerning societal perception, along with its potential negative ramifications, creates a gap in comprehending technology adoption processes. Failure to address social acceptability risks overlooking crucial aspects of technology acceptance and use, including concerns related to stigmatization and negative judgments from others, as well as considerations of admiration and pride. Hence, investigating social acceptability warrants significant attention within the realm of technology acceptance models.

2.2 Social Acceptability of Technology

Social acceptability can be conceptualized as the "absence of social disapproval" [13]. Understanding societal perceptions that form the basis for social approval or disapproval is crucial, as they shape not only the observations of technology use in others but also influence individuals' expectations of societal endorsement or rejection upon embracing new technologies themselves [14, 15]. Social acceptability is recognized as a reciprocal and context-dependent process, not merely an isolated occurrence. Using public transportation, including AEBs, per definition involves participation in a social space. Drawing on the theory of impression management, which posits that all public behaviors are performative acts, it becomes evident that individuals craft their identities not solely for self-expression but also to navigate social interactions—and thus are constantly monitoring the impression they might be making on other people with their behavior [16]. This multifaceted concept of social acceptability is further explored by Montero et al. (2010), who break it down into two distinct aspects: the user's social acceptance, which influences the individual's personal perception, and the observer's social acceptance, which impacts how the individual's use of technology is perceived by others [17]. This interplay suggests that the presence and responses of onlookers can profoundly influence users, potentially leading to feelings of discomfort, embarrassment, or a reduced propensity to engage with the technology [18, 19]. This underscores the importance of considering both individual and societal perspectives in the adoption of new technologies.

Research by Nordhoff et al. (2019) highlights the impact of firsthand experience with AEBs [20]. Specifically, individuals who had used an AEB were found to express highly positive sentiments towards it, indicating that direct interaction frequently leads to favorable perceptions. Marsden et al. (2019) explored user perceptions of autonomous electric delivery vehicles within a living lab environment [21]. These vehicles are designed for last-mile delivery services, equipped with parcel lockers accessible to recipients through a dedicated app. Participants predominantly viewed these autonomous delivery vehicles as innovative, environmentally sustainable, and intriguing.

The discussion on the social acceptability of AEBs requires understanding potential gender differences. Research indicates that vehicle innovations often align with the preferences of specific demographics, revealing a gender bias [22]. Men and women may perceive autonomous vehicles and their potential use differently. This discrepancy suggests that AEBs might not equally cater to women's needs as they do to men's, reflecting and potentially reinforcing societal stereotypes. Gender differences in the perception of AEBs also need to be considered in the light of stereotype threat [23], i.e., societal stereotypes influencing individual behaviors to conform, thus further complicating the gendered dynamics of AEB acceptance. While some studies did not find any effect of gender on the acceptance of AEBs [2], others have demonstrated that both acceptance and the intent to use autonomous vehicles can indeed differ according to gender: For autonomous delivery vehicles, female participants were more critical about these vehicles compared to male participants and exhibited a slightly lower inclination towards using them in a living lab environment [24]. Moreover, some UTAUT2 constructs were only significant for women: Women valued the enjoyment of using AEBs as a transport option higher than men, considered social factors more, and considered potential risk when using autonomous delivery vehicles more than men [24]. Hohenberger et al. (2016) identified a difference between women and men in their willingness to use autonomous vehicles, which was linked to their emotional reactions towards them [25]. These findings underscore the importance of recognizing and addressing the diverse perceptions and needs shaped by the social construction of gender in the development and social acceptability of autonomous transport.

The referenced studies evaluated factors affecting the societal perceptions and acceptability of AEBs and similar vehicles. This aspect of acceptance, often labeled as "acceptability" or "social acceptance" [7, 14, 26], shows a significant relationship with the SCM dimensions [9]. This connection underlines the importance of social perceptions in technology adoption, highlighting the intricate interplay between societal norms and technological advancements. The subsequent sections will elaborate how the SCM framework elucidates this relationship.

2.3 The Stereotype Content Model

Social perception can generally be distilled into two fundamental dimensions [9, 27], firstly competence, agency, "getting ahead", and secondly warmth, communion, "getting along". These dimensions are foundational in shaping our understanding and judgments of others within social contexts, offering a structured approach to analyze and interpret social acceptability and biases [27, 28]. Warmth assesses benevolence and likability, while competence gauges capability and independence. The framework of the SCM [9] shows how high ratings on both dimensions are for admired ingroups, while outgroups are often seen ambivalently. For instance, "rich people" may be seen as competent but cold, evoking envy, while groups like housewives may be viewed as warm but incompetent, eliciting pity. Groups perceived as both cold and incompetent, like homeless people, are despised, while those warm and competent, like one's ingroup, are admired, leading to "in-group favoritism." Groups classified similarly in these dimensions, i.e., belonging to the same cluster, tend to evoke similar emotional and behavioral responses [9, 28]. The SCM predicts stereotypical status based on four combinations: paternalistic (high warmth, low competence), admirable (high warmth, high competence), contemptuous (low warmth, low competence), and envious (low warmth, high competence) stereotypes (see Fig. 1).

The SCM was successfully used to understand the social perception of different user groups depending on their use of different mobile devices as well as establishing that devices are perceived stereotypically by themselves (e.g., smart glasses, VR headsets, tablets, EEG headsets) [26]. Regarding the relationship between the SCM and the social acceptance of mobile devices, Schwind et al. (2019) demonstrated that the intention and capability to achieve a goal are influenced by the interaction between the user and the device. User-device combinations suggesting low status and vying for similar resources are perceived as disdainful and, consequently, were less socially accepted. Conversely, user-device combinations suggesting high status and minimal competition are esteemed and, therefore, socially embraced [26]. Frischknecht's (2021) research examining the overall perception of autonomous technical systems also references the SCM: Her findings indicate that as technical autonomy levels rise, there are changes in how competence and agency are attributed, whereas attributions of warmth and experience remain relatively stable [29].

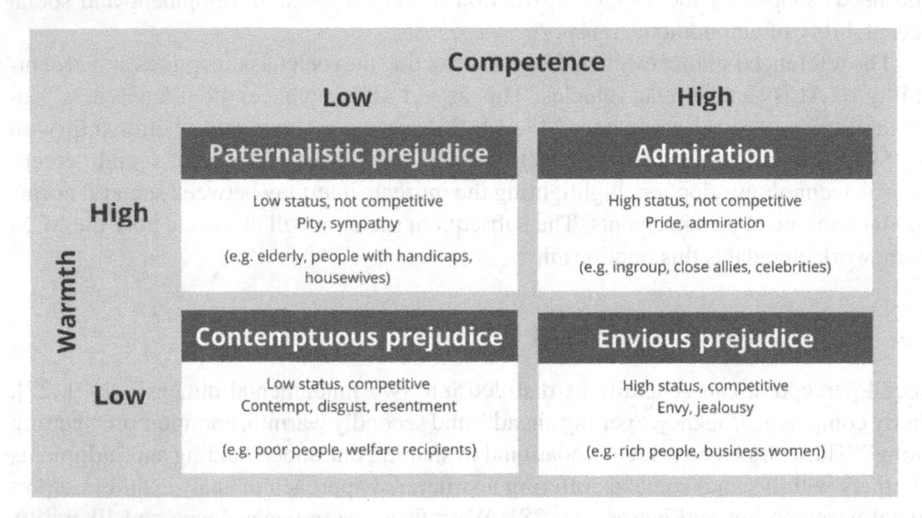

Fig. 1. The Stereotype Content Model (SCM), figure adapted from Fiske et al. [9]

In our exploration of AEBs, the synthesis of related work on social acceptability and the SCM facilitates the formulation of hypotheses centered around the generally positive perceptions of AEB users, the impact of in-group effects from AEB utilization, and the role of gender stereotypes.

Generally Positive Perception of AEB Users. Previous research shows that users of autonomous delivery vehicles generally received high competence ratings compared to people who did not use autonomous delivery vehicles [7]. In addition, using public transportation instead of individual modes of transportation such as cars, is usually considered as more environmentally friendly [30], which can be linked to positive social aspects, as environmentalists are considered rather warm [26]. Furthermore, as people in technical professions rate high on competence [9, 31], we expect the people who use autonomous public transportation will be seen as more competent than people who reject it. Thus, we propose that people who use AEBs are seen as more competent and warmer than people who reject these (H1 and H2).

In-Group Effects of AEB Use. People tend to reduce their stereotypical notions when they actually have contact with a group of people or other source of stereotypes [32]. Moreover, mere categorizations as having something in common with a group of people can create in-group effects, i.e., having something in common with an otherwise unspecified group may lead to a more favorable perception of these group members [33]. Thus, we propose that participants, who have previous experiences with using AEBs themselves, will rate people who use AEBs as warmer and more competent than people without experience (H3, H4) and rate people who reject using AEBs as comparably less competent and less warm (H5, H6). Similar effects are also assumed for participants who express a high intention to use AEBs, as this can been seen as an indicator of a positive attitude, therefore, we presume the same pattern of warmth and competence ratings for participants with a high intention to use AEBs compared to those with low intentions (H7, H8).

Influence of Gender Stereotypes. Based on the gendered perception of autonomous vehicles and the influences of societal stereotypes on individual behavior to conform we expect men to express more positive attitudes than women in our study, i.e., men rate people who use AEBs as more competent and warmer than women do (H9), and they rate people who reject AEBs as less competent and warm compared to women (H10) (Table 1).

Table 1. Hypotheses regarding the Stereotype Content Model in terms of AEBs.

Generally positive perception of AEB users	
H1	Users of AEBs are rated as more competent than people who reject using autonomous vehicles
H2	Users of AEBs are rated as warmer than people who reject using autonomous vehicles
In-group effects of AEB use	
H3	Participants who have prior experience with using AEBs will rate people using AEBs as warmer compared to participants who don't have prior experience
H4	Participants who have prior experience with using AEBs rate people using AEBs as more competent compared to participants who don't have prior experience
H5	Participants who have prior experience with using AEBs rate people who reject using AEBs as less warm compared to participants, who don't have prior experience
H6	Participants who have prior experience with using AEBs rate people who reject using AEBs as less competent compared to participants who don't have prior experience
H7	Participants scoring high on their intention to use AEBs in the future rate people rejecting AEBs as less warm than participants scoring low on the intention to use the shuttle
H8	Participants scoring high on their intention to use AEBs in the future see people rejecting AEBs as less competent than people scoring low on the intention to use the shuttle
Influence of gender stereotypes	
H9	Male participants rate people who use AEBs as more competent and warmer compared to female participants
H10	Male participants rate people who reject AEBs as less competent and warm compared to female participants

3 Methodological Approach

We conducted our study on the SCM among both users and non-users in a mid-sized town in Germany, where AEBs have been undergoing a pilot phase in public transport since August 2022.

3.1 Measurement of Social Perception

In line with previous research of measuring the SCM [7, 34], we used "warmherzig" (warm), "freundlich" (good-natured), and "sympathisch" (likeable) for assessing the warmth dimension; and we used "kompetent" (competent), "eigenständig" (independent), and "konkurrenzfähig" (competitive) for the competence dimension. Participants rated both "people using the AEB" and "people rejecting the AEB" on a five-point Likert-scale (1 = strongly agree to 5 = strongly disagree), additionally participants could select an alternative "I don't know" response option.

3.2 Procedure: Survey, Sample and Tools for Analysis

The online survey ran from December 2022 to February 2023 on the Keyingress software platform. For recruiting study participants, we used social media, flyers, and directly asked AEB users. Altogether, $N = 223$ people participated in the study. Of those, $N = 56$ participants already used an AEB. The sample was relatively young (72% in between 16–40 years) with gender being equally represented ($N = 84$ male, $N = 72$ female, and $N = 3$ diverse participants). However, due to missing values and drop-out rates we could only use a subsample ($N = 166$) for the analysis. In addition, as respondents could also select the "I don't know" category only around 50% of the responses could be used in the final SCM analysis. For data analysis, we used IBM Statistics SPSS Version 26 and R 4.2.2. We centered all variables and effect-coded nominal, dichotomous variables.

4 Results

The analysis shows significant differences in the perception of people's competence ($t = -7.08, p = .000, df = 80$) and warmth ($t = -6.312, p = .000, df = 73$) based on their use or rejection of the shuttle. Thus, the data obtained are consistent with H1 and H2, suggesting that people using AEBs are seen as warmer ($M = 1.99$) and more competent ($M = 1.97$) than people rejecting it ($M = 3.29$ for warmth and $M = 3.36$ for competence) (see Fig. 2).

This difference is even more evident when comparing the responses of participants with and without experience of using the AEB: Participants with experience ($M = 1{,}67$) perceived people who use AEBs more positively on the warmth dimension than participants without experience ($M = 2{,}14$, $t = -2{,}29$, $p = .026$, $df = 55{,}028$). We did not find a significant difference on the competence dimension ($t = -1{,}516$, $p = .135$, $df = 59{,}6$) (see Fig. 3). Thus, we could confirm H3, but not H4.

People's perception of people rejecting AEBs does not differ on the warmth ($t = 1{,}717, p = .0943, df = 37$) and competence ($t = 1.608, p = .116, df = 39{,}45$) dimension

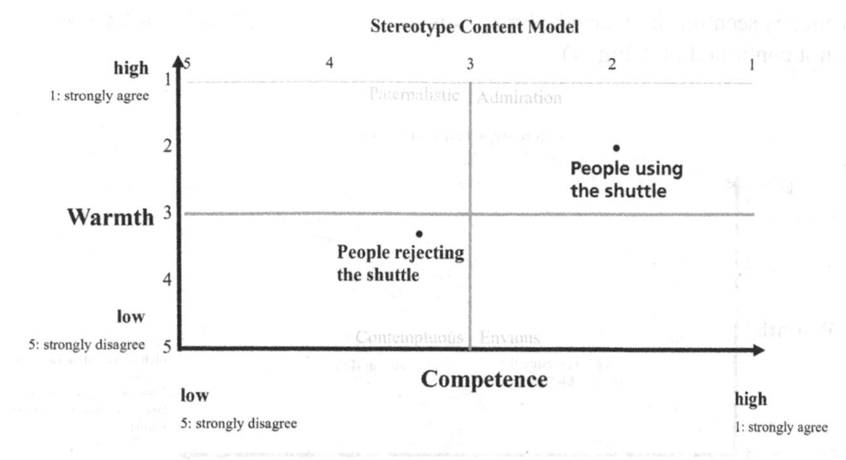

Fig. 2. Ratings of people using and rejecting AEBs.

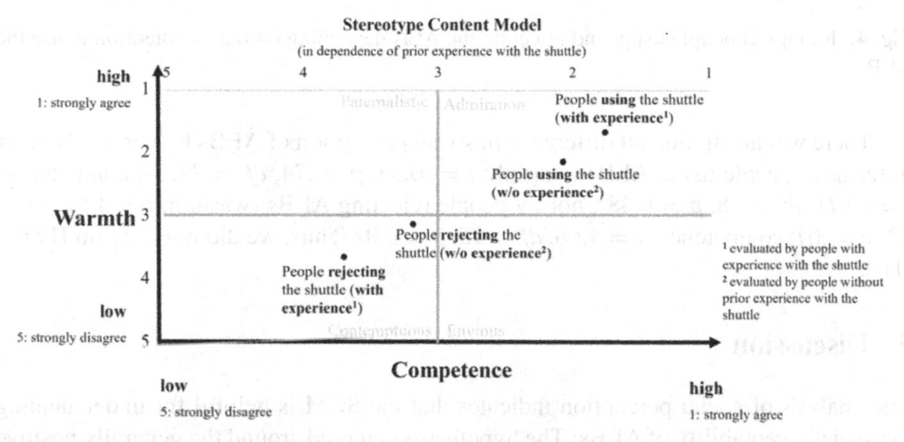

Fig. 3. Ratings of people using and rejecting the AEB depending on raters' experience with / without the AEB.

based on their experience with AEBs. Only descriptively, and marginally significant, we found that people rejecting AEBs are perceived as less warm by people with experience with the AEB than without (see Fig. 3). Therefore, we could neither confirm H5, nor H6.

When comparing the judgments of people scoring high vs. low on their intention to use AEBs in future, the perception of people using AEBs differs on both dimensions (warmth: t = -3,24, p = .01, df = 9,07; competence: t = -3,96, p = .0011, df = 15,84). People rejecting AEBs are seen as less competent by people scoring high on their intention to use AEBs in the future than people scoring low on the intention to use AEBs (t = 2,211, p = .045, df = 13,513); therefore, H7 is confirmed. No significant difference is seen on the warmth dimension ($t = 1.28$, $p = .23$, $df = 8{,}74$); therefore, H8 is not confirmed (see Fig. 4).

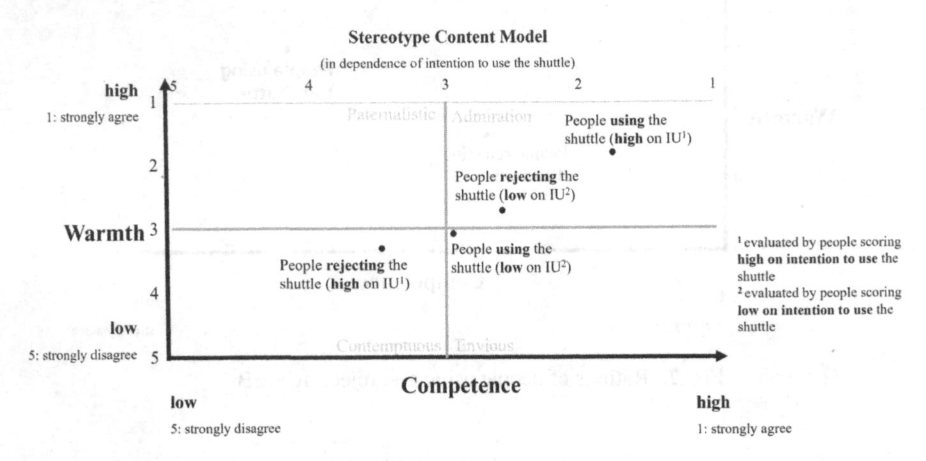

Fig. 4. Ratings of people using and rejecting the AEB depending on raters' intention to use the AEB.

There was no significant difference in social perception of AEBs by gender. Neither in terms of people using AEBs (warmth: $t = -0{,}33$, $p = .74$, $df = 77{,}18$; competence: $t = -0{,}71$, $df = 88$, $p = 0.48$), nor by people rejecting AEBs (warmth: $t = 1{,}82$, $df = 72$, $p = .07$; competence: $t = 1{,}66$, $df = 75$, $p = .10$). Thus, we did not confirm H9 nor H10.

5 Discussion

The analysis of social perception indicates that the SCM is helpful for understanding the social acceptability of AEBs: The hypotheses centered around the generally positive perceptions of AEB users are significant. Of the six hypotheses regarding the impact of in-group effects from AEB use, only two are significant. The hypotheses focusing on the role of gender stereotypes are not significant.

First, the hypotheses regarding a more positive perception on the SCM dimensions of people who use AEBs compared to people who reject using AEBs is significant: We found that people using AEBs were considered as more competent (H1) and warmer (H2) than people who reject using AEBs.

Second, the hypotheses on the impact of in-group effects from AEB use, i.e., favoritism based on perceived similarity regarding the use of AEBs, yielded mixed results. Participants with prior experience using AEBs rated users of AEBs as warmer (H3), but—in contrast to our hypothesis—not as more competent than participants without prior experience (H4). Contrary to our hypotheses, prior experience with AEBs did not affect the perceived warmth (H5) and competence (H6) of people who reject AEBs. Furthermore, participants' intention to use AEBs in the future did not affect their perception of warmth of those who rejected AEBs (H7). In line with our hypothesis, a higher intention to use was linked to lower competence perceptions of them (H8). Third, in contrast to our hypothesis, we did not find any effect of gender (H9 and H10).

The findings on the social perceptions of AEB users align partially with prior studies on the social acceptability of AEBs. Similar to our findings, Pröbster and Marsden (2023) identified a trend of attributing higher competence to users of AEBs compared to non-users [7]. This study, alongside our current research, did not observe significant gender-based differences in perceptions (unlike [25], which found gender differences), suggesting that factors beyond gender play a role in shaping social perceptions of AEB users.

As expected, we found a gain in warmth of users compared to people who reject using AEBs. We ascribe this to the positive stereotypical perceptions of public autonomous transportation, which can be associated with environmentally friendly behavior [35]. The perception of users of AEB seems to be seen as more communal and more positive as the perception of users of autonomous delivery vehicles, which deliver goods to separate households and were associated with a loss of warmth for some user groups [7]. Referring to the implications of Schwind et al. (2016) regarding the SCM, people using AEBs may also been seen as more device dependent (as this is considered public transport) than people using autonomous delivery vehicles, resulting in lower status but higher warmth [26].

We expected in-group favoritism by participants with prior experience regarding people using AEBs compared to people without prior experience, but this effect was only observed for warmth and not for competence ratings. Arguably, as experienced participants were aware of the ease of use of AEBs, while participants without experience may overestimate the difficulty, competence ratings of users may have converged for both groups.

As prior experiences did not affect warmth and competence ratings of people who reject AEBs, we assume that experience lead to a certain in-group favoritism of users, but without the devaluation of the out-group. The experience with AEBs may not have been entirely positive for some members of the experienced group and thus, rejecting AEBs may not be seen as negatively as we assumed. But this contradicts the overall positive perceptions of people who use AEBs compared to those who reject them. Another possible explanation may be that people are aware that a lot of people have yet to try out AEBs as they are still rare in Germany, and their rejection can be attributed to their inexperience. Considering the AEBs are means of public transportation, rejecting them might be partly interpreted as not being dependent on public transport and thus, not lead to a loss in competence. However, further studies seem necessary to gain a better understanding of inter-group perceptions of users and non-users of AEBs.

The future intention of participants to utilize AEBs did not significantly influence their perception of warmth towards individuals who rejected AEBs. However, a greater inclination to adopt AEBs was associated with a diminished perception of competence in non-users. This observation aligns with our initial hypotheses and suggests that the intention to use AEBs serves as a marker for the strength of identification with being an AEB user. Consequently, the devaluation of competence in out-group members, i.e., those who do not use AEBs, may be attributed to a heightened identification with AEB usage. However, we need to consider that the cell frequencies for these comparisons were very low. Thus, further research could investigate how the intention to use AEBs affects the perceptions of these groups.

In general, the findings indicate a positive perception of AEBs, with potential users being rated high in both warmth and competence according to the SCM. Compared to individuals who reject AEBs, users are perceived as even more competent and warm. According to the SCM, higher warmth means that the person is more likely to try to "get along" rather than "get ahead" [27], so the elevated warmth ratings associated with AEB users suggest lower competitiveness, implying that AEB users may be perceived as less inclined to be rivals for the same resources compared to users of other technological devices such as autonomous delivery vehicles [7] and AR-glasses [26].

Why is it Important to Go Beyond Acceptance Models? In our study, we discovered that the SCM captures concepts pertinent to the acceptance of AEBs. This suggests that a comprehensive understanding of people's attitudes and acceptance of AEBs necessitates the integration of automatic, intuitive processes captured by the SCM in addition to rational considerations accounted for in acceptance models like UTAUT or TAM as done in prior studies.

- The SCM illuminates automatic processes that occur swiftly and often subconsciously, offering insights into individuals' ingrained attitudes and assumptions that may not be consciously considered, especially in the early stages of innovation adoption. Conversely, acceptance models incorporate various factors such as expected performance, ease of use, and other rational arguments.
- By encompassing social situations and considering the looking glass self—the idea that our self-concept is shaped by how we believe others perceive us—the SCM extends the scope of existing acceptance models like UTAUT or TAM, providing deeper insights into the implicit attitudes that influence social perceptions of AEBs beyond mere factors of rational acceptance.
- The process of rating individuals' warmth and competence based on their usage or rejection of AEBs, akin to the formation of stereotypes in everyday life, appears to occur more automatically compared to the deliberate evaluation of factors like performance expectancy or effort expectancy, particularly when detailed technological knowledge is lacking. Moreover, underlying psychological processes such as fear, trust, and sympathy may be subconsciously influenced by stereotype ratings, bypassing the need for active consideration as required by acceptance models' item batteries.

Social-desirability bias, a potential response bias prevalent in acceptance models due to participants' inclination towards providing socially acceptable responses, might skew results towards more positive outcomes. Conversely, uncertainties surrounding technology may not be openly acknowledged within acceptance models, whereas in the SCM construct, the direct inquiry into attitudes towards the technology circumvents this issue.

- Drawing from Kahneman's 2-systems [36], which posits that intentions, including stereotypes and emotions, are often formulated in 'System 1'—the intuitive, automatic system—before being processed by the more rational 'System 2'. This suggests that stereotypical attitudes may prelude rational decision-making, highlighting the importance of understanding the role of stereotypes in shaping perceptions of technology.
- Nevertheless, it is important to acknowledge that stereotypes are, among other factors, influenced by prior experiences and knowledge, as also indicated by Frischknecht (2021). Similarly, factors used in an acceptance model such as UTAUT show varying effects over time [37]. Thus, further research is needed to comprehensively understand the evolving acceptance and usage processes of AEBs over time.

6 Limitations

Our study is subject to several limitations that must be considered when interpreting the results. First, our sample was relatively specific, consisting mainly of young participants. Furthermore, the AEB tour commencing at a science museum may have primed participants with an inherent openness towards technological innovations, potentially biasing their responses towards AEBs. Moreover, our dataset exhibited a notable rate of missing values, with the possibility of missing not at random (MNAR) data. While we addressed this issue by excluding participants with missing values from the analysis, it is important to acknowledge that this may have introduced further bias into our findings.

7 Conclusion

Our research highlights the role of social acceptability of (semi-)autonomous public transport: The integration of a sociopsychological aspects through the SCM showed that perceptions differ between users and non-users and that the measurement of stereotypes reveals insights to understand people's attitudes and acceptance of AEBs. Our findings underscore a need for further research to find ways to assess potential users' acceptance of automated public transport, which should incorporate all relevant aspects of acceptance, including social acceptance measured by the stereotypical notions of a technology and of the people using it.

Acknowledgments. This work has been partially funded by the European Commission in the program "HORIZON.4.2—Reforming and enhancing the European R&I System" under the topic "HORIZONWIDERA-2022-ERA-01–80—Living Lab for gender-responsive innovation" as part of the project "GILL—Gendered Innovation Living Labs", grant agreement ID 101094812. The responsibility for all content supplied lies with the authors.

Disclosure of Interests. The authors have no competing interests to declare that are relevant to the content of this article.

References

1. Korkmaz, H., Fidanoglu, A., Ozcelik, S., Okumus, A.: User acceptance of autonomous public transport systems: extended UTAUT2 model. J. Public Transp. **24**, 100013 (2022)
2. Madigan, R., Louw, T., Wilbrink, M., Schieben, A., Merat, N.: What influences the decision to use automated public transport? using UTAUT to understand public acceptance of automated road transport systems. Transport. Res. F: Traffic Psychol. Behav. **50**, 55–64 (2017)
3. Iclodean, C., Cordos, N., Varga, B.O.: Autonomous shuttle bus for public transportation: a review. Energies **13**(11), 2917 (2020)
4. Herrenkind, B., Brendel, A.B., Nastjuk, I., Greve, M., Kolbe, L.M.: Investigating end-user acceptance of autonomous electric buses to accelerate diffusion. Transp. Res. Part D: Transp. Environ. **74**, 255–276 (2019)
5. Pigeon, C., Alauzet, A., Paire-Ficout, L.: Factors of acceptability, acceptance and usage for non-rail autonomous public transport vehicles: a systematic literature review. Transport. Res. F: Traffic Psychol. Behav. **81**, 251–270 (2021)
6. Marsden, N., Pröbster, M.: The social perception of autonomous delivery vehicles. Scholarly Community Encyclopedia (2024). https://encyclopedia.pub/entry/53514
7. Pröbster, M., Marsden, N.: The social perception of autonomous delivery vehicles based on the stereotype content model. Sustainability **15**(6), 5194 (2023)
8. Cuddy, A., Fiske, S., Glick, P.: Warmth and competence as universal dimensions of social perception: the stereotype content model and the BIAS map. Adv. Exp. Soc. Psychol. **40**, 61–149 (2008). https://doi.org/10.1016/S0065-2601(07)00002-0
9. Fiske, S., Cuddy, A., Glick, P., Xu, J.: A model of (often mixed) stereotype content: competence and warmth respectively follow from perceived status and competition. J. Pers. Soc. Psychol. **82**(6), 878–902 (2002). https://doi.org/10.1037/0022-3514.82.6.878
10. Madigan, R., et al.: Acceptance of automated road transport systems (ARTS): an adaptation of the UTAUT model. Transport. Res. Procedia **14**, 2217–2226 (2016)
11. Venkatesh, V., Morris, M.G., Davis, G.B., Davis, F.D.: User acceptance of information technology: toward a unified view. MIS Q. 425–478 (2003). https://doi.org/10.2307/30036540
12. Nielsen, J.: Usability Engineering. en. Google-Books-ID: DBOowF7LqIQC. Elsevier (1994)
13. VandenBos, G.R.: APA dictionary of psychology. American Psychological Association (2007)
14. Sehrt, J., Braams, B., Henze, N., Schwind, V.: Social acceptability in context: stereotypical perception of shape, body location, and usage of wearable devices. Big Data Cogn. Comput. **6**(4), 100 (2022). https://doi.org/10.3390/bdcc6040100
15. Profita, H.P.: Designing wearable computing technology for acceptability and accessibility. ACM SIGACCESS Accessibil. Comput. **114**, 44–48 (2016). https://doi.org/10.1145/2904092.2904101
16. Goffman, E.: The presentation of self in everyday life. Doubleday, Garden City (1959)
17. C. S. Montero, J. Alexander, M. T. Marshall, and S. Subramanian, "Would you do that? Understanding social acceptance of gestural interfaces," in *Proceedings of the 12th international conference on Human computer interaction with mobile devices and services*, 2010, pp. 275–278, doi: https://doi.org/10.1145/1851600.1851647
18. Y.-T. Hsieh, A. Jylhä, V. Orso, L. Gamberini, and G. Jacucci, "Designing a willing-to-use-in-public hand gestural interaction technique for smart glasses," in *Proceedings of the 2016 CHI Conference on Human Factors in Computing Systems*, 2016, pp. 4203–4215, doi: https://doi.org/10.1145/2858036.2858436
19. Lucero, A., Vetek, A.: NotifEye: using interactive glasses to deal with notifications while walking in public. In: Proceedings of the 11th Conference on Advances in Computer Entertainment Technology, pp. 1–10 (2014). https://doi.org/10.1145/2663806.2663824

20. Nordhoff, S., de Winter, J., Payre, W., Van Arem, B., Happee, R.: What impressions do users have after a ride in an automated shuttle? an interview study. Transport. Res. F: Traffic Psychol. Behav. **63**, 252–269 (2019). https://doi.org/10.1016/j.trf.2019.04.009
21. Marsden, N., Dierolf, N., Herling, C.: HCI research for responsible innovation: a living-lab approach to designing an automated transport system for last mile logistics (2019)
22. Christensen, H.R., Breengaard, M.H., Levin, L.: Gender Smart Mobility: Concepts, Methods, and Practices. Routledge, Abingdon (2023)
23. Inzlicht, M., Schmader, T.: Stereotype Threat: Theory, Process, and Application. Oxford University Press, Oxford (2012)
24. Kapser, S., Abdelrahman, M., Bernecker, T.: Autonomous delivery vehicles to fight the spread of Covid-19–How do men and women differ in their acceptance? Transport. Res. Part A: Policy Pract. **148**, 183–198 (2021). https://doi.org/10.1016/j.tra.2021.02.020
25. Hohenberger, C., Spörrle, M., Welpe, I.M.: How and why do men and women differ in their willingness to use automated cars? the influence of emotions across different age groups. Transport. Res. Part A: Policy Pract. **94**, 374–385 (2016). https://doi.org/10.1016/j.tra.2016.09.022
26. Schwind, V., Deierlein, N., Poguntke, R., Henze, N.: Understanding the social acceptability of mobile devices using the stereotype content model. In: Proceedings of the 2019 CHI Conference on Human Factors in Computing Systems, pp. 1–12 (2019). https://doi.org/10.1145/3290605.3300591
27. Abele, A.E., Ellemers, N., Fiske, S.T., Koch, A., Yzerbyt, V.: Navigating the social world: toward an integrated framework for evaluating self, individuals, and groups. Psychol. Rev. **128**(2), 290 (2021). https://doi.org/10.1037/rev0000262
28. Fiske, S., Cuddy, A., Glick, P.: Universal dimensions of social cognition: warmth and competence. Trends Cogn. Sci. **11**(2), 77–83 (2007). https://doi.org/10.1016/j.tics.2006.11.005
29. Frischknecht, R.: A social cognition perspective on autonomous technology. Comput. Hum. Behav. **122**, 106815 (2021). https://doi.org/10.1016/j.chb.2021.106815
30. Holland, S.P., Mansur, E.T., Muller, N.Z., Yates, A.J.: The environmental benefits of transportation electrification: urban buses. Energy Policy **148**, 111921 (2021)
31. Cuddy, A.J., Fiske, S.T., Glick, P.: The BIAS map: behaviors from intergroup affect and stereotypes. J. Pers. Soc. Psychol. **92**(4), 631 (2007). https://doi.org/10.1037/0022-3514.92.4.631
32. Pettigrew, T.F.: Intergroup contact theory. Annu. Rev. Psychol. **49**(1), 65–85 (1998)
33. Tajfel, H., Turner, J.C.: The social identity theory of intergroup behavior. In: Austin, W.G., Worchel, S. (eds.) Psychology of Intergroup Relations, pp. 7–24. Nelson-Hall Publishers, Chicago (1986)
34. Asbrock, F.: Stereotypes of social groups in Germany in terms of warmth and competence. Social Psychol. **41**(2), 76 (2010). https://doi.org/10.1027/1864-9335/a000011
35. Cai, L., Yuen, K.F., Wang, X.: Explore public acceptance of autonomous buses: an integrated model of UTAUT, TTF and trust. Travel Behav. Soc. **31**, 120–130 (2023)
36. Kahneman, D.: Thinking, Fast and Slow. Macmillan, New York (2011)
37. Blut, M., Chong, A., Tsiga, Z., Venkatesh, V.: Meta-analysis of the unified theory of acceptance and use of technology (UTAUT): challenging its validity and charting a research agenda in the red ocean. J. Assoc. Inf. Syst. (2021)

The Development of Human-Centered Design in Public Transportation: A Literature Review

Chikita Rini Lengkong(✉), Cindy Mayas, Heidi Krömker, and Matthias Hirth

Technische Universität Ilmenau, Ilmenau, Germany
{chikita.lengkong,cindy.mayas,heidi.kroemker,matthias.hirth}@tu-ilmenau.de

Abstract. Public transportation aims to meet the mobility needs of heterogeneous target groups. Therefore, the importance of human-centered methods for the further development of transport and related information services is growing. This paper analyzes the development of the application of human-centered design in the context of public transportation. According to the Preferred Reporting Items for Systematic Reviews and Meta-Analyses (PRISMA), a literature review in ACM Digital Library, IEEE Xplore, Web of Science, and Scopus database is conducted. The analysis represents publications from 32 countries or territories globally. As a result, 106 publications with the year of publication between 2004 and 2023 show the growing importance of human-centered methods, such as interviews, observations, surveys, user studies, and prototyping for different application areas in public transport. The results reveal a trend towards quantitative and computational methods in addition to the qualitative user-centered methods for networking information systems related to public transport. This trend is observed alongside the growing implementation of automation systems in vehicles.

Keywords: human-centered design · user-centered design · public transportation · literature review

1 Introduction

The growing challenges of urbanization and climate change underline the importance of prioritizing public transport. Public transport is associated with significant contributions to the local community and living conditions, particularly in terms of economic and environmental aspects [93]. Economically, efficient public transport systems can stimulate economic growth by improving accessibility and connectivity. As the infrastructures of public transportation advance, locals are able to mobilize for diverse purposes, spanning business, employment, recreation, education, and various activities, thereby enhancing their overall quality of life [101]. From an environmental perspective, public transport plays a crucial role in reducing the carbon footprint associated with the usage of private

H. Krömker (Ed.): HCII 2024, LNCS 14733, pp. 40–62, 2024.
https://doi.org/10.1007/978-3-031-60480-5_3

vehicles [39]. As a result, public transport helps establish healthier and more sustainable living environments for the locals.

The adaptability and focus on end-user needs make Human-Centered Design (HCD) an effective approach for establishing innovative and effective solutions in the public transportation system. Several studies have shown that HCD has been widely used in various disciplines, including healthcare [76], workspace [13], manufacturing industry [50], tourism technology [91], and urban concept [16]. The implementation of HCD aims to design products or services that align with user needs and desires by taking the interaction of human beings with the environment into consideration [6]. The application of HCD encourages people to use public transportation as their preferred mode of travel, which is in line with sustainability and the development of a climate-neutral city. Additionally, applying HCD results in the improvement of community engagement and promotes a mutual relationship between all stakeholders involved, including passengers, drivers, operators, and city planners.

First analysis research in the area of transportation systems shows the rising importance of User Experience (UX) and usability in transportation [98]. While the analysis focuses on UX and usability in driving from 2000 to 2019, a significant research gap exists concerning the application of the HCD principle in public transportation services for a diverse target group. Investigating the development of HCD in public transportation systems is crucial for addressing usability challenges and enhancing passenger experience. Prioritizing research in this area leads to the design of more user-friendly public transportation systems, thereby potentially increasing public usage and overall satisfaction among stakeholders in public transportation. Considering the importance of HCD in public transportation and the advantage of literature review, this paper conducted a literature review of the scientific publications related to HCD in public transportation from 2004 onwards that are collected from four journal databases. Besides the literature review, country performance, trends and methods, and keywords co-occurrence analysis were also conducted based on the result from the literature review. The paper presents a collection of literature connecting HCD principles with public transportation. Additionally, key methodologies and emerging trends in the development of HCD in public transportation were identified, providing researchers with a resource for further exploration and for conducting future studies in this field.

This study is guided by the following research questions, which serve as the foundation for the investigation and analysis:

RQ1 How has the application of HCD developed in the context of public transportation?

RQ2 What are the methodologies and topics of HCD that have been applied in the context of public transportation?

The remainder of this paper is structured as follows. Section 2 discusses related works. Section 3 describes the methods used for the literature search, filtering, and analysis. The results of the literature review are presented in Sect. 4. Finally, Sect. 5 concludes the paper.

2 Human-Centered-Design Public Transportation

The consideration of user needs in different development processes is discussed in the scientific literature under the terms "human-centered" and "user-centered".

In addition to the use of the system by users, the term "human-centered" also takes into account its impact on stakeholders who are not directly involved in its use [21]. This includes drivers and passengers of public transport, staff in control centers, and economic and political decision-makers, who may be indirectly influenced by or exert influence upon the systems themselves.

The term "user-centered" primarily refers to development processes in which the development team takes the needs of the users into account [1]. For this purpose, methods for direct user participation, such as interviews, observation, and usability tests are used. Additionally, indirect consideration of users' needs through expert evaluations is also put into practice.

Since both Human-Centered Design (HCD) and User-Centered Design (UCD) aims at an improved adaptation of the public transport system to different usage situations and perspectives, both terms are considered in the analysis of this paper. Moreover, the application of HCD/UCD in public transportation offers numerous advantages to diverse stakeholders involved in the transportation system. Firstly, for commuters, HCD/UCD ensures that transportation systems are designed and built to meet the commuters' specific needs and preferences [27,79,119]. By prioritizing UX, HCD/UCD can enhance overall comfort and satisfaction among passengers, leading to an increase in the number of usages [5]. Secondly, for transportation authorities and policymakers, HCD/UCD provides valuable insights into people's travel behavior and facilitates planners and policymakers to identify areas with great potential for improvement, enabling them to design more efficient, accessible, and sustainable transportation solutions [39]. Furthermore, HCD/UCD facilitates insights into enhancing safety and security for public transportation crews, operators, and drivers, thereby contributing to an overall improvement in transportation services [25]. Moreover, HCD/UCD promotes inclusiveness by considering the diverse needs of different demographic groups, including the elderly [18,67], disabled [20,104], and economically disadvantaged individuals [2]. In summary, by prioritizing the needs and experiences of users, HCD/UCD in public transportation can benefit commuters, transportation operators, urban planners, and local communities as a whole, leading to the improvement of the public transportation system. Through a literature review, insight into existing research is provided, allowing the identification of trends and methodologies that can be used for guiding future studies.

3 Method

This section details how the the literature review was conducted. First, the search terms and the search process is described. Second, filter criteria for identifying relevant papers are presented. Finally, specific evaluation questions are formulated to assess each publication.

3.1 Exclusion, Inclusion and Search Process

A systematic literature review is conducted to provide an analysis of existing research in the field of HCD/UCD in public transportation between 2004–2023. The data is gathered from four databases: ACM Digital Library, IEEE Xplore, Web of Science, and Scopus by using the specified keywords for the search criteria. These databases are chosen for their reliability and extensive coverage of academic literature, consequently supporting the consistency of the literature review on the development of HCD/UCD in public transportation. The search strings are selected to accommodate a wide range of transportation modes as well as using different ways of spelling HCD and UCD. In particular, the following search strings are used.

Search string 1 "human centered design" OR "human centred design" OR "human-centered design" OR "human-centred design" OR "user centered design" OR "user centred design" OR "user-centered design" OR "user-centred design"

Search string 2 "public transportation" OR "mass transportation" OR "mass transit" OR "bus line" OR "trolley bus" OR "bus rapid transit" OR "bus service" OR "shuttle bus" OR "light rail" OR "LRT" OR "minibus" OR "metro" OR "tram" OR "rail rapid transit" OR "railway" OR "regional rail" OR "interurban" OR "rapid transit" OR "subway" OR "urban transportation service" OR "mono rail" OR "rail road" OR "intercity rail" OR "high speed rail" OR "commuter rail" OR "urban rail transit" OR "metrolink" OR "people mover" OR "street car"

The resulting number of publications is summarized in Table 1. Before analyzing the retrieved publications, the following criteria were applied: written in English, scientific articles, and published after the year 2000. Although the data retrieval spans from 2000, the results on HCD/UCD in public transportation are observed starting from 2004. Hence, the analysis will continue from that year onwards.

Table 1. Systematic literature search results on HCD/UCD in public transportation.

Databases	Results	Last retrieved date
ACM Digital Library	544	11.01.2024
IEEE Xplore	557	11.01.2024
Scopus	83	11.01.2024
Web of Science	18	11.01.2024

3.2 Search Filters

Both empirical and theoretical studies are included in the analysis. According to previous studies, the utilization of the Preferred Reporting Items for Systematic Reviews and Meta-Analyses (PRISMA) methodology is linked to a more comprehensive reporting of systematic reviews [72, 74]. Therefore, this literature review relies on a systematic approach by using this methodology. The PRISMA flow chart is applied to provide visualization on how the papers are identified, screened, and included in the review based on the criteria that have been selected. After the publications were retrieved, a manual screening was performed to find suitable studies that cover the scope of HCD/UCD in public transportation. The following filters were used as a guideline for the screening:

1st Filter Removing duplicate records
2nd Filter Selecting the article by reading the title and abstract whether the article fits the study
3rd Filter Reviewing the full papers for eligibility
Final result Analyzing the publications

The search filters in literature reviews have a function to clarify the search process, enhancing relevance and efficiency. This helps to focus on studies aligning research objectives while preventing information overload. Transparent documentation of the filtering process enables the review process to be replicated for future studies with the same interest in the method.

3.3 Quality Evaluation

The publications that are eligible to be evaluated were read and categorized. The following questions were asked to obtain a quality index for each publication:
EQ1 Is the concept of "human-centered design" or "user-centered design" the core of the study?
EQ2 Is "public transportation" or "means of public transportation" the core subject of the study?
EQ3 Are users or participants directly involved in the study?

For each question, papers that provide satisfactory answers are marked as "yes", while those that do not meet the criteria are marked as "no". Papers failing to meet the criteria for **EQ 2** are still included in the analysis if public transportation is referenced as a secondary research topic. Similarly, for **EQ 3**, papers addressing HCD/UCD are included in the analysis despite the absence of direct human involvement.

4 Result

This section summarizes the results of the literature review. The section first details on the results of the literature search and categorizes the relevant papers according to the evaluation questions introduced in Sect. 3.

To answer the two research questions **RQ 1** and **RQ 2** operationalization is applied. An analysis is performed based on the countries of origin and publication years to answer research question **RQ 1**. To address **RQ 2**, each publication is coded based on eight categories of methodology and a keyword co-occurrence analysis is conducted.

4.1 Preferred Reporting Items for Systematic Reviews and Meta-Analyses

Figure 1 shows that a total of 1202 scientific works were retrieved from the four databases and reviewed systematically. A total of 23 duplicate articles were removed from the screening process, leaving 1179 articles to be reviewed. During the records screening process, the titles and abstracts of these 1179 articles were reviewed. Among them, 137 articles were identified as proceedings, while 911 articles were identified as not thematic to the field of HCD/UCD and public transportation and therefore, are excluded from the next step.

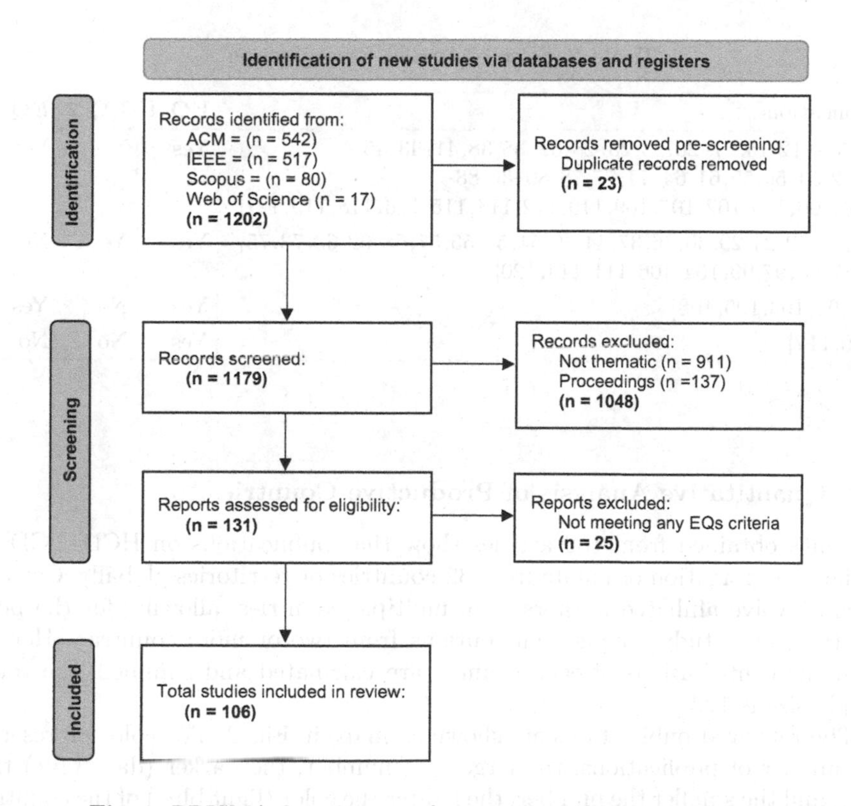

Fig. 1. PRISMA flow diagram for systematic literature review.

After title and abstract screening, the full text of 131 articles was screened based on the quality evaluation questions. Following the full text screening, 25 articles did not meet any of the three quality evaluation criteria **EQ 1**, **EQ 2**, and **EQ 3**, and were excluded from the final review. At the end of the screening process, 106 publications were eligible to be analyzed.

Table 2 highlights distinct patterns in publication. A total of 70 publications met the criteria **EQ 1**, **EQ 2**, and **EQ 3**, indicating a main focus on public transportation and user participation. Furthermore, 29 publications met **EQ 1** and **EQ 2** indicating a focus on the core concepts of human-centered design and the primary subject of public transportation. However, these publications are absent from direct involvement or engagement with users or participants. This absence suggests that direct input from individuals was not integrated when data collection was conducted. A total of 5 publications only fulfilled **EQ 1** and **EQ 3**, suggesting a prioritization of human-centered design and other fields that also put interest in public transportation. Finally, 2 publications met only **EQ 1**, referring to a narrower focus on human-centered design's principle without involving data collection from users or participants and without having public transportation as their main and only research.

Table 2. Selected studies for evaluation.

Publication	**EQ 1**	**EQ 2**	**EQ 3**
[2–5,8–12,15,17–20,23,25–28,32–35,38,41–43,45–48,52,54,56,59,61,64–71,77,78,80–86,88–90,94,96,100,102,107,109,110,112,113,115,116,118,119,121]	Yes	Yes	Yes
[7,14,22,24,29,30,36,37,44,49,51,53,55,57,58,62,63,73,75,79,87,95,97,99,104,106,111,114,120]	Yes	Yes	No
[31,92,103,105,108]	Yes	No	Yes
[40,117]	Yes	No	No

4.2 Quantitative Analysis of Productive Countries

The data obtained from 106 articles show that publications on HCD/UCD in public transportation originate from 32 countries or territories globally. Certain papers involve affiliated authors from multiple countries, allowing for the possibility of one study having contributions from two or more countries. Hence, when the contributions of each country are calculated and summed, the total sample size is 122.

The retrieved publications are shown as maps in Fig. 2. The color represents the number of publications, the larger the number, the darker (dark blue) the color, and the smaller the number, the lighter the color (light blue) of the country. The map visualizes a concentration of research in North America, Europe, and East Asia. It suggests a notable level of interest and investment in this field of research within these regions. While the concentration of research activity in the map shows the significant development of HCD/UCD in public transportation

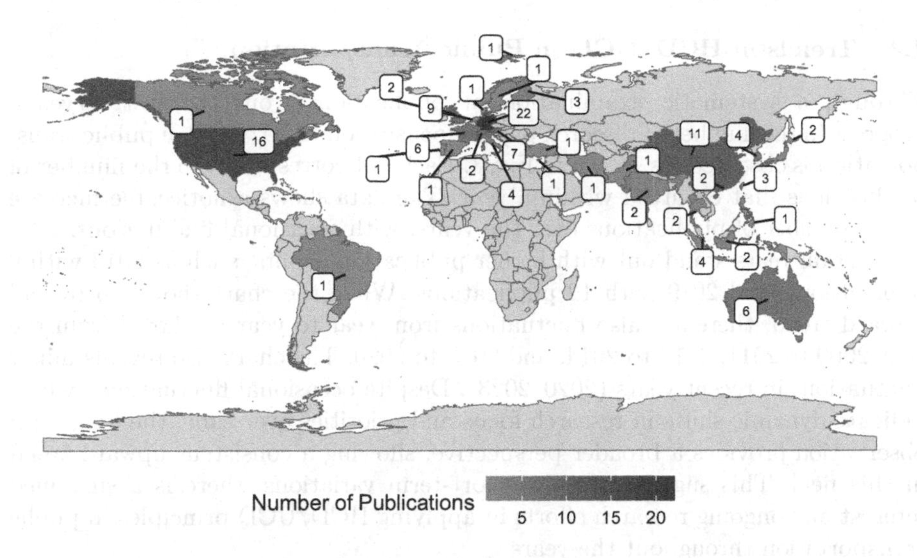

Fig. 2. Publications per country.

Table 3. Top 10 productive countries in HCD/UCD in public transportation, 2004-2023.

Rank	Countries/Territories	Region	Quantity
1	Germany	Europe	22
2	United States	North America	16
3	China	Asia	11
4	United Kingdom	Europe	9
5	Austria	Europe	7
6	France	Europe	6
7	Australia	Oceania	6
8	Singapore	Asia	4
9	South Korea	Asia	4
10	Italy	Europe	4

in these regions, it also presents an opportunity for collaboration with countries that contribute to the most publications.

Table 3 lists the top 10 most productive countries, collectively contributing 89 samples, representing 73% of the considered publications. The leading countries in publications are Germany, the United States, and China. Germany and the United States represent 18% and 13% of the total numbers, followed by China with 9%. The United Kingdom and Austria are fourth and fifth, respectively covering 7% and 6%. Lastly, the average percentage of papers published in countries between the 6th and 10th places is 4%.

4.3 Trends in HCD/UCD in Public Transportation

Through a systematic examination of publications from the last 20 years, progress in the number of research papers focusing on HCD/UCD in public transportation is observed, as shown in Fig. 3. Each dot corresponds to the number of publications that occurred within a year. The data shows a noticeable increase in the number of publications over the years, with occasional fluctuations.

Certain years stand out with higher publication counts, such as 2013 with 9 publications, and 2019 with 12 publications. While the chart shows an overall upward trend, there are also fluctuations from year to year as shown from the year 2009 to 2011, 2012 to 2014, and 2018 to 2020. The chart also reveals minor fluctuations in recent years (2020–2023). Despite occasional fluctuations, which indicate dynamic shifts in research focus and priorities over time, the long-term observation provides a broader perspective, showing a consistent upward trend in this field. This suggests despite short-term variations, there is a sustained interest and ongoing research efforts in applying HCD/UCD principles in public transportation throughout the years.

Figure 4 shows the methods applied in HCD/UCD publications in public transportation spanning from 2004 to 2023. The analysis reveals a diverse selection of methodologies applied to investigate HCD/UCD in public transportation, for example, interviews and focus groups, observation, content analysis, survey and questionnaire, user studies and usability testings, prototype and design, computational experiment and simulation, and other methods. The publications are categorized based on all methodologies applied, with the majority of publications using more than one methodology. These methods are grouped for analysis; for instance, various types of observations including bird's-eye observation, shadow observation, user or self-observation, video observation, and field observation are categorized as observation. Prototyping, design, system design, architectural design, design engineering, modeling frameworks, mock-ups, and pilot testing are displayed as prototypes and designs. Additionally, user journeys, personas, and heuristic approaches are categorized under user studies, while data mining, big data analysis, and crowdsourcing are classified as other methods.

The number of publications operating interviews and focus groups, surveys, observations, and prototypes and designs as methods generally show an increas-

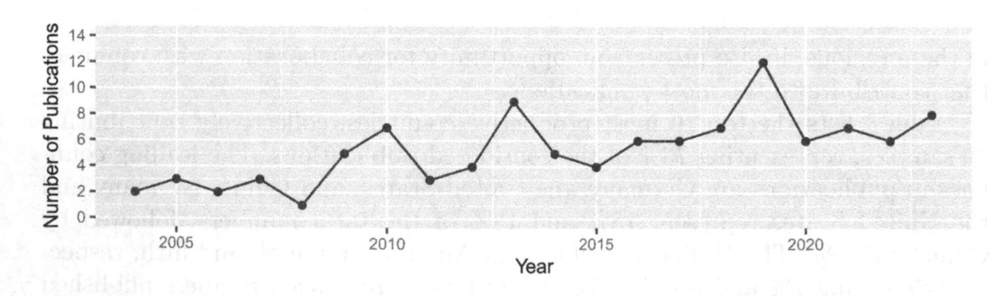

Fig. 3. Number of publications on the topic of HCD/UCD in public transportation.

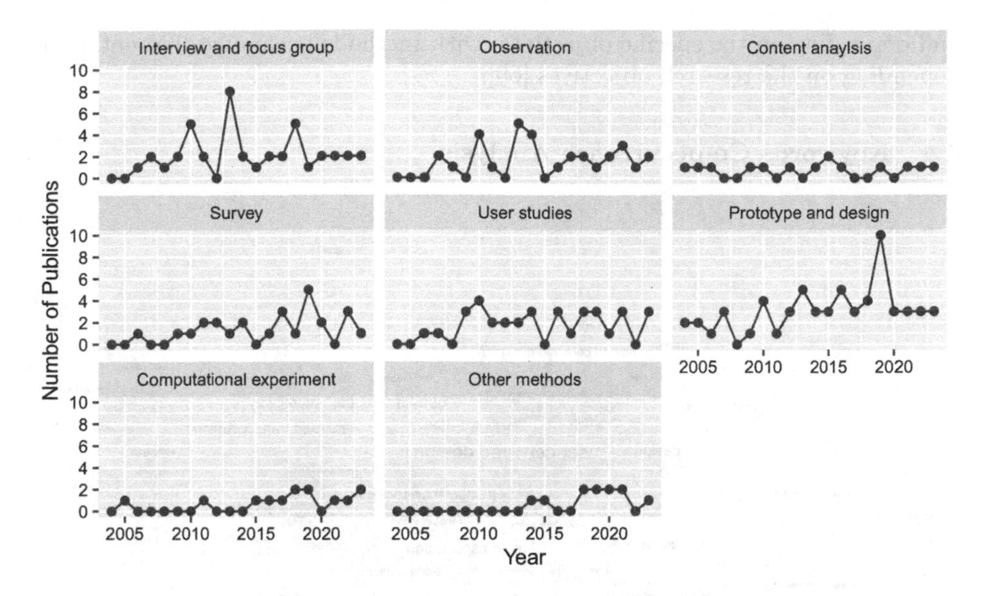

Fig. 4. Methods trends in HCD/UCD in public transportation.

ing trend over time, with periodic fluctuations. Prototype and design emerged as the most frequent methods used in HCD/UCD in public transportation, covering a total sample size of 52 with the highest spike in 2019 with 10 publications. Following closely behind is the interview and focus group, with a total sample size of 42. Although it is one of the most popular methods overall, its usage experienced a decrease after 2013 before showing a slight increase in later years. User studies and observation followed with 35 and 31 samples, respectively. Despite fluctuations, observations, and user studies remain a significant method for studying user behavior and user experience in public transportation. The widespread application of interviews, user studies, and observation also suggests a reliance on qualitative data collection methods, which prioritize a direct engagement with users or participants. Moreover, the survey accounted for the fifth most applied method, with a sample size of 26. Although not consistently the most applied methods in comparison to the others, surveys show increased usage in the past five years, featuring in 11 publications, following closely behind prototype and design as the highest in this period, which appeared in 22 samples. These results mark a potential shift towards quantitative data collection methods. Content analysis as a method that provides insight into textual data rounded out number six with 14 samples. Lastly, computational experiments and other methods had the lowest sample size, with 13 and 11, respectively. Despite not being consistently applied, these categories represent a diverse range of research approaches, highlighting the multidisciplinary nature of HCD/UCD in public transportation research. This finding indicates that there is a growing diversity of methods used within the scope of HCD/UCD in public transportation and depending on the specific objectives, with methods becoming differentiated depending on the research objectives [60].

4.4 Keywords Co-occurrence Analysis

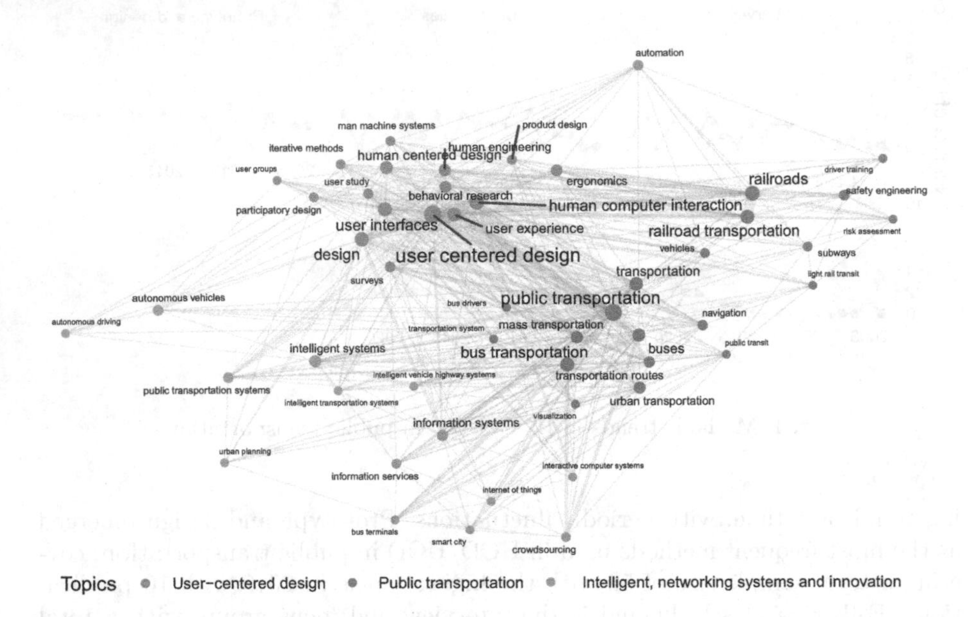

Fig. 5. The keywords co-occurrence network of HCD/UCD in public transportation.

The keywords chosen for a publication serve as indicators of the main contents, while the frequency of the keywords in academic literature specifies the significance of other associated topics within a particular field [98]. This method enables an understanding of the content of individual publications and the broader thematic trends within HCD/UCD in public transportation.

In this research, the normalization of keywords, analysis, and visualization was conducted using R Studio. Keyword normalization was performed to ensure consistent treatment of the singular and plural forms, synonyms, and orthography. For example, "rails", "railway", "railways", and "railroad" were normalized to "railroads". To understand the links between various topics in the literature works, a selection condition was set, which means that keywords with a frequency of at least 4 appearances were recognized for the keywords co-occurence analysis. The outcome revealed a total of 51 qualified keywords and 462 links. In the following, we focus on this keyword corpus.

The keyword co-occurrence networks in this field is shown in Fig. 5. In Fig. 5, each node represents a keyword, with the size of the node corresponding to the number of occurrences of the keyword and the edges connecting the nodes represent a co-occurrence of the two keywords in a publication. The weight of the edges, visualized by its size, represents the frequency of the co-occurrence of the keywords, i.e., the larger the edge between two keywords, the more often they appear together in different publications.

Table 4. Top 5 keywords of each cluster.

Clusters	Keywords	Occurrences	Ranks	Strength
User-centered design				
	user centered design	35	1	116
	human computer interaction	17	3	65
	design	15	5	61
	user interfaces	14	8	52
	user experience	13	9	58
Public transportation				
	public transportation	28	2	106
	railroads	17	4	60
	bus transportation	14	6	64
	railroad transportation	14	7	52
	transportation	12	10	48
Intelligent, network system, and innovation				
	intelligent systems	9	14	40
	information systems	7	19	32
	automation	6	21	24
	autonomous vehicles	6	22	18
	public transportation systems	6	23	25

The qualified keywords were manually clustered into three main topics "user center design", "public transportation" and "intelligent, network system, and innovation". The clusters are also visualized in Fig. 5, identified by the color blue for "user center design", green for "public transportation", and red for "intelligent, network system, and innovation". The top 5 keywords with the highest frequency for each cluster are listed in Table 4. The occurrences indicate how often a keyword was used in different papers. Ordering the keywords by their number of occurrences results in rank shown in the table, with the most frequently used keyword being assigned to rank 1. Finally, the strength of a keyword is based on its weighted node degree in the network.

Keywords affiliating to the clusters blue and green are among the top 10 keywords with the highest occurrences, in which "user centered design" is the most frequent keyword, with 35 occurrences, followed by "public transportation" and "human computer interaction" with 28 and 17 occurrences respectively. The keywords "railroads" and "design" are also in the top 5 of high-frequently used keywords. This suggests that these keywords have the most direct connections to other indexes and function as a bridge in the research system and each of the top 5 keywords listed for each cluster could be considered as a focus within the field to a certain extent. The result also shows that "railroads", "bus transportation", and "railroad transportation" are high in occurrence. This dominance indicates the relevance and importance of these mode of transportation in public transportation system.

The keywords "intelligent system", "information systems", "automation", "autonomous vehicles", and "public transportation systems" emerge as the most frequently occurring terms in the "intelligent network systems and innovation" domain. Moreover, the frequent appearance of automation and autonomous vehicles in this cluster indicates a growing interest or focus on within the context of HCD/UCD in public transportation.

5 Conclusion

Based on the dataset of 1202 documents obtained from ACM, IEEE Xplore, Scopus, and Web of Science in the field of HCD/UCD in public transportation, this study conducted a literature analysis on 106 publications. The following analyses were conducted to address the research questions about the methodologies and topics of HCD/UCD in public transportation and the development of its application.

To conclude, the research addressed two key questions. In response to **RQ 1**, a trend showing a recognition of this topic over time is observed, as illustrated by the growing number of publications in the last 20 years. The findings also present evidence indicating that the research topic is widely spread across diverse geographic locations and research disciplines. Furthermore, a set of relevant papers was listed, serving as a resource for further exploration and deepening the understanding of HCD/UCD application in public transportation. To address **RQ 2**, the findings indicate a growth in the application of quantitative and computational method for networking information systems in public transportation. Moreover, prototype and design are found to be the most applied method in the field of HCD/UCD in public transportation. Furthermore, growing implementation of automation systems in vehicles is also observed. Additionally, in public transportation modes, railroads and bus transportation as are among the top keywords that appeared.

Despite efforts to systematically review and evaluate relevant publications, the exclusion of other journal databases and unpublished studies may have limited the analysis. Consequently, the coverage and quantity of publications were restricted. In future studies, it is recommended to include relevant publications from additional databases, such as Springer, JSTOR, and Taylor & Francis Online. Additionally, future studies can integrate qualitative data analysis to have a depth of understanding of user experiences and preferences within the context of HCD/UCD in public transportation.

Acknowledgments. Parts of this work were funded by the German Federal Ministry for Digital and Transport (BMDV) grant number 45AVF3004G within the project OeV-

LeitmotiF-KI and by the German Federal Ministry of Education and Research (BMBF) grant number 21INVI2302 within the project NetÖV.

Disclosure of Interests. The authors have no competing interests to declare that are relevant to the content of this article.

References

1. User-centered design (2016). https://www.interaction-design.org/literature/topics/user-centered-design, (Accessed 11 February 2024)
2. Kalejaiye G.B., Orefice H.R., Moura T.A., Bafutto M., Carvalho M.M.: Poster abstract: Frugal crowd sensing for bus arrival time prediction in developing regions. In: 2017 IEEE/ACM Second International Conference on Internet-of-Things Design and Implementation (IoTDI), pp. 355–356 (2017)
3. Aguiar A., Cruz Nunes F.M., Fernandes Silva M.J., Silva P.A., Elias D.: Leveraging electronic ticketing to provide personalized navigation in a public transport network. IEEE Trans. Intell. Trans. Syst. **13**(1), 213–220 (2012). https://doi.org/10.1109/TITS.2011.2167612
4. Anderson, R.E., et al.: Experiences with a transportation information system that uses only GPS and SMS. In: Proceedings of the 4th ACM/IEEE International Conference on Information and Communication Technologies and Development, ICTD 2010. Association for Computing Machinery, New York (2010). https://doi.org/10.1145/2369220.2369223
5. Beul-Leusmann, S., Jakobs, E.-M., Ziefle, M.: User-centered design of passenger information systems. In: IEEE International Professonal Communication 2013 Conference, pp. 1–8 (2013). https://doi.org/10.1109/IPCC.2013.6623931
6. van der Bijl-Brouwer, M., Dorst, K.: Advancing the strategic impact of human-centred design. Des. Stud. **53**, 1–23 (2017). https://doi.org/10.1016/j.destud.2017.06.003
7. Bouargane, L., Cherkaoui, A.: Towards an explicative model of human cognitive process in a hidden hazardous situation and a cognitive ergonomics intervention in railway environment. In: 2015 International Conference on Industrial Engineering and Systems Management (IESM), pp. 968–976 (2016). https://doi.org/10.1109/IESM.2015.7380272
8. Brandenburg E., Kozachek D., Konkol K., Woelfel C., Geiger A., Stark R.: How pedestrians perceive autonomous buses: evaluating visual signals. In: 2021 IEEE 2nd International Conference on Human-Machine Systems (ICHMS), pp. 1–4 (2021). https://doi.org/10.1109/ICHMS53169.2021.9582644
9. Brandenburger, N., Naumann, A.: Towards remote supervision and recovery of automated railway systems: The staff's changing contribution to system resilience. In: Int. Conf. Intell. Rail Transp., ICIRT. Institute of Electrical and Electronics Engineers Inc. (2018). https://doi.org/10.1109/ICIRT.2018.8641576
10. Brossard, A., Abed, M., Kolski, C., Uster, G.: User modelling: The consideration of the experience of time during journeys in public transportation. In: Proceedings of the 6th International Conference on Mobile Technology, Application & Systems, Mobility 2009, Association for Computing Machinery, New York (2009). https://doi.org/10.1145/1710035.1710079

11. Cai T., Hong X., Lin Z., Zhao W.: Design and implementation of intelligent bus-stop boards system. In: 2023 4th International Conference on Computer Engineering and Application (ICCEA), pp. 396–399 (2023). https://doi.org/10.1109/ICCEA58433.2023.10135435
12. Campbell, M., Bennett, C., Bonnar, C., Borning, A.: Where's my bus stop? Supporting independence of blind transit riders withStopInfo. In: Proceedings of the 16th International ACM SIGACCESS Conference on Computers & Accessibility, ASSETS 2014, pp. 11–18. Association for Computing Machinery, New York (2014). https://doi.org/10.1145/2661334.2661378
13. Cawood, T., et al.: Creating the optimal workspace for hospital staff using human centred design. Intern. Med. J. **46**(7), 840–845 (2016). https://doi.org/10.1111/imj.13124
14. Chang, S.J., Hsu, G.Y., Huang, S.J.: Location-aware mobile transportation information service. In: 2005 2nd Asia Pacific Conference on Mobile Technology, Applications and Systems, pp. 5 (2005). https://doi.org/10.1109/MTAS.2005.207220
15. Chen, Y.C., Lai, H.H.: Computer linguistic automatic ticket-selling system for the blind: The perspective of human-sense characteristics. J. Converg. Inf. Technol. **6**(2), 41–49 (2011). https://doi.org/10.4156/jcit.vol6.issue2.5
16. Choi, J., Kim, S., Min, D., Lee, D., Kim, S.: Human-centered designs, characteristics of urban streets, and pedestrian perceptions. J. Adv. Transp. **50**(1), 120–137 (2016). https://doi.org/10.1002/atr.1323
17. Cornet, H., Stadler, S., Kong, P., Marinkovic, G., Frenkler, F., Sathikh, P.: User-centred design of autonomous mobility for public transportation in Singapore. In: Antoniou C., Busch F., Moeckel R., Wulfhorst G., Pfertner M. (eds.) Transp. Res. Procedia, vol. 41, pp. 191–203. Elsevier B.V. (2019). https://doi.org/10.1016/j.trpro.2019.09.038
18. Deb, D., Deb, T., Sharma, M.: Redesign of a railway coach for safe and independent travel of elderly. In: Tanveer, M., Pachori, R.B. (eds.) Machine Intelligence and Signal Analysis. AISC, vol. 748, pp. 147–156. Springer, Singapore (2019). https://doi.org/10.1007/978-981-13-0923-6_13
19. Degen, H., Guillen, G., Schmidt, H.: Effectiveness and cost-benefit ratio of weekly user group sessions. In: Marcus, A., Wang, W. (eds.) HCII 2019. LNCS, vol. 11583, pp. 208–221. Springer, Cham (2019). https://doi.org/10.1007/978-3-030-23570-3_16
20. Detjen, H., Schneegass, S., Geisler, S., Kun, A., Sundar, V.: An emergent design framework for accessible and inclusive future mobility. In: Proceedings of the 14th International Conference on Automotive User Interfaces and Interactive Vehicular Applications, AutomotiveUI 2022, pp. 1–12, Association for Computing Machinery, New York (2022). https://doi.org/10.1145/3543174.3546087
21. DIN Deutsches Institut für Normung e. V.: DIN EN ISO 9241-210:2020-03, Ergonomie der Mensch-System-Interaktion - Teil 210: Menschzentrierte Gestaltung interaktiver Systeme; Deutsche Fassung. Tech. rep., Beuth Verlag GmbH (2020). https://doi.org/10.31030/3104744, https://www.beuth.de/de/-/-/313017070
22. Duszek, R.: The warsaw subway signage system. Inform. Des. J. **12**(3), 181–186 (2004). https://doi.org/10.1075/idjdd.12.3.02dus
23. Ekşioğlu, M.: User experience design of a prototype kiosk: A case for the Istanbul public transportation system. Inter. J. Hum.-Comput. Interact. **32**(10), 802–813 (2016). https://doi.org/10.1080/10447318.2016.1199179

24. Elliott A.C.: Human factors for railway signalling and control systems. In: IET Professional Development Course on Railway Signalling and Control Systems (RSCS 2012), pp. 236–248 (2012). https://doi.org/10.1049/ic.2012.0056
25. Esmaeili, L., Hashemi G, S.: Rural intelligent public transportation system design: applying the design for re-engineering of transportation eCommerce system in Iran. Inter. J. Inform. Technol. Syst. Approach **8**(1), 1–27 (2015). https://doi.org/10.4018/IJITSA.2015010101
26. Foong, P.S., Diaz, V.J., Houssian, A.R., Huse, A., Jamsri, P.: EventStream: integrated transit information system. In: CHI 2007 Extended Abstracts on Human Factors in Computing Systems, CHI EA 2007 pp. 2061–2066. Association for Computing Machinery, New York (2007). https://doi.org/10.1145/1240866.1240950
27. Foth, M., Schroeter, R.: Enhancing the experience of public transport users with urban screens and mobile applications. In: Proceedings of the 14th International Academic MindTrek Conference: Envisioning Future Media Environments, MindTrek 2010, pp. 33–40. Association for Computing Machinery, New York (2010). https://doi.org/10.1145/1930488.1930496
28. Fox, C., Oliveira, L., Kirkwood, L., Cain, R.: Understanding users' behaviours in relation to concentrated boarding: Implications for rail infrastructure and technology. In: El Souri M., Gao J., Keates S. (eds.) Advances in Transdisciplinary Engineering, vol. 6, pp. 120–125. IOS Press BV (2017). https://doi.org/10.3233/978-1-61499-792-4-120
29. Franke, I.S., Köhler, A., Elze, A.: Control the autonomous! - user interfaces for monitoring and dispatching autonomous vehicles. In: 2021 IEEE International Intelligent Transportation Systems Conference (ITSC), pp. 2292–2297 (2021). https://doi.org/10.1109/ITSC48978.2021.9564635
30. Frizziero, L., Donnici, G., Galiè, G., Pala, G., Pilla, M., Zamagna, E.: QFD and SDE methods applied to autonomous minibus redesign and an innovative Mobile Charging System (MBS). Inventions **8**(1) (2023). https://doi.org/10.3390/inventions8010001
31. Gabrielli, S., Maimone, R.: Are change strategies affecting users' transportation choices? In: Proceedings of the Biannual Conference of the Italian Chapter of SIGCHI, CHItaly 2013, Association for Computing Machinery, New York (2013). https://doi.org/10.1145/2499149.2499155
32. Gasté Y., Gentes A.: "Place" and "non-place": A model for the strategic design of place-centered services. Bell Labs Tech. J. **17**(4), 21–36 (2013). https://doi.org/10.1002/bltj.21572
33. Gnauer C., et al.: Knowledge based training derived from risk evaluation concerning failure mode, effects and criticality analysis in autonomous railway systems. In: 2021 5th International Conference on System Reliability and Safety (ICSRS), pp. 47–52 (2021). https://doi.org/10.1109/ICSRS53853.2021.9660726
34. Gran, B., Rindahl, G., Sarshar, S., Lunde-Hanssen, L.: Design including RAMS and security for new traffic control centers for the norwegian railway. In: Beer M., Zio E. (eds.) Proceedings of European Safety and Reliability Conference, ESREL. pp. 1757 1763. Research Publishing Services (2020). https://doi.org/10.3850/978-981-11-2724-3_0245-cd
35. Grippenkoven, J., Naumann, A., Bhattacharyya, A., Lemmer, K.: Form follows vision - user-centred interface design for rail traffic controllerś workplaces. In: IFAC Proceedings Vol. (IFAC-PapersOnline), vol. 3, pp. 89–93. IFAC Secretariat (2013). https://doi.org/10.3182/20131111-3-KR-2043.00023, issue: PART 1 Journal Abbreviation: IFAC Proc. Vol. (IFAC-PapersOnline)

36. He Y., Ni J., Niu B., Li F., Shen X.: Privacy-preserving ride clustering for customized-bus sharing: A fog-assisted approach. In: 2018 16th International Symposium on Modeling and Optimization in Mobile, Ad Hoc, and Wireless Networks (WiOpt), pp. 1–8 (2018). https://doi.org/10.23919/WIOPT.2018.8362850
37. Heming Zhang, D. Chen: Developing a multidisciplinary approach for concurrent engineering and collaborative design. In: 8th International Conference on Computer Supported Cooperative Work in Design, vol. 2, pp. 449–454 (2004). https://doi.org/10.1109/CACWD.2004.1349230
38. Hildén, E., Väänänen, K., Chistov, P.: Travel experience toolkit: bus-specific tools for digital service design. In: Proceedings of the 17th International Conference on Mobile and Ubiquitous Multimedia, MUM 2018, pp. 193–197. Association for Computing Machinery, New York (2018). https://doi.org/10.1145/3282894.3282916
39. Jiang, Y., Zhou, Z., Liu, C.: The impact of public transportation on carbon emissions: a panel quantile analysis based on Chinese provincial data. Environ. Sci. Pollut. Res. **26**(4), 4000–4012 (2019). https://doi.org/10.1007/s11356-018-3921-y
40. Jiang, S., Ferreira, J., Gonzalez, M.C.: Activity-based human mobility patterns inferred from mobile phone data: a case study of Singapore. IEEE Trans. Big Data **3**(2), 208–219 (2017). https://doi.org/10.1109/TBDATA.2016.2631141
41. Kang, Y., Park, S., Kim, K., Kim, H.: Development of interactive map-based ui for subway ticketing kiosk system. In: Proc. IADIS International Conference Interfaces and Human Computer Interaction, IHCI, Proc. IADIS International Conference on Game and Entertainment Technologies, Part MCCSIS, pp. 319–322 (2010)
42. Karvonen, H., Aaltonen, I., Wahlström, M., Salo, L., Savioja, P., Norros, L.: Unraveling metro train driver's work: Challenges in automation concept. In: Proceedings of the 28th Annual European Conference on Cognitive Ergonomics, ECCE 2010, pp. 233–240. Association for Computing Machinery, New York (2010). https://doi.org/10.1145/1962300.1962349
43. Kim, J.E., Bessho, M., Koshizuka, N., Sakamura, K.: Enhancing public transit accessibility for the visually impaired using IoT and open data infrastructures. In: Proceedings of the First International Conference on IoT in Urban Space, URB-IOT 2014, pp. 80–86. ICST (Institute for Computer Sciences, Social-Informatics and Telecommunications Engineering), Brussels, BEL (2014). https://doi.org/10.4108/icst.urb-iot.2014.257263
44. Kim B., et al.: Suggestion on the practical human robot interaction design for the autonomous driving disinfection robot. In: 2021 21st International Conference on Control, Automation and Systems (ICCAS), pp. 2068–2073 (2021). https://doi.org/10.23919/ICCAS52745.2021.9649781
45. Kjeldskov, J., Andersen, E., Hedegaard, L.: Designing and evaluating buster: An indexical mobile travel planner for public transportation. In: Proceedings of the 19th Australasian Conference on Computer-Human Interaction: Entertaining User Interfaces, OZCHI 2007, pp. 25–28. Association for Computing Machinery, New York (2007). https://doi.org/10.1145/1324892.1324897
46. Klein, P., Cetin, N.: User-centered development of real-time dispatcher clients. In: IEEE International Professional Communication Conference, pp. 210–213 (2010). https://doi.org/10.1109/IPCC.2010.5530011
47. Kulkarni, M., et al.: BusStopCV: a real-time ai assistant for labeling bus stop accessibility features in streetscape imagery. In: Proceedings of the 25th International ACM SIGACCESS Conference on Computers and Accessibility, ASSETS

2023, Association for Computing Machinery, New York (2023). https://doi.org/10.1145/3597638.3614481
48. Kuys, J., Melles, G., Al Mahmud, A., Thompson-Whiteside, S., Kuys, B.: Human centred design considerations for the development of sustainable public transportation in malaysia. Appli. Sci. (Switzerland) **12**(23) (2022). https://doi.org/10.3390/app122312493
49. Kuys, J.O., Melles, G., Thompson-Whiteside, S., Kapoor, A.: Scoping the human-centred design led 2020 vision for Malaysia electric buses (EV). In: 18th Asia Pacific Automotive Engineering Conference, APAC 2015 (March 2015). https://doi.org/10.4271/2015-01-0031
50. Kuys, J., Al Mahmud, A., Kuys, B.: A case study of university-industry collaboration for sustainable furniture design. Sustainability **13**(19) (2021). https://doi.org/10.3390/su131910915
51. Kieu, L.M., Bhaskar, A., Chung, E.: Passenger segmentation using smart card data. IEEE Trans. Intell. Transp. Syst. **16**(3), 1537–1548 (2015). https://doi.org/10.1109/TITS.2014.2368998
52. Leitner, M., Subasi, O., Höller, N., Geven, A., Tscheligi, M.: User requirement analysis for a railway ticketing portal with emphasis on semantic accessibility for older users. In: Proceedings of the 2009 International Cross-Disciplinary Conference on Web Accessibililty (W4A), pp. 114–122. W4A 2009. Association for Computing Machinery, New York (2009). https://doi.org/10.1145/1535654.1535683
53. Li, Q., Liu, L., Zhang, W., Liu, J., Bai, G.: Evaluating regional rail transit safety: a matter-element analysis method. In: 2020 IEEE 23rd International Conference on Intelligent Transportation Systems (ITSC), pp. 1–6 (2020). https://doi.org/10.1109/ITSC45102.2020.9294374
54. Libreros, J.A., Mayas, C., Hirth, M.: Recommender systems in continuing professional education for public transport: challenges of a human-centered design. In: Adjunct Proceedings of the 31st ACM Conference on User Modeling, Adaptation and Personalization, UMAP 2023, pp. 331–336. Adjunct. Association for Computing Machinery, New York (2023). https://doi.org/10.1145/3563359.3596995
55. Lu, Y., Wu, H., Liu, X.: Chen P,: TourSense: a framework for tourist identification and analytics using transport data. IEEE Trans. Knowl. Data Eng. **31**(12), 2407–2422 (2019). https://doi.org/10.1109/TKDE.2019.2894131
56. Magrini, L., Nati, M., Panizzi, E.: RMob - a mobile app for real time information in urban transportation. In: Proceedings of the International Working Conference on Advanced Visual Interfaces, AVI 2012 pp. 776–777. Association for Computing Machinery, New York (2012). https://doi.org/10.1145/2254556.2254709
57. Martí, P., Jordán, J., González Arrieta, A., Julian, V.: A survey on demand-responsive transportation for rural and interurban mobility. Intern. J. Interact. Multimedia Artifi. Intell. **8**(3), 43–54 (2023). https://doi.org/10.9781/ijimai.2023.07.010
58. Mathew, A.P.: Using the environment as an interactive interface to motivate positive behavior change in a subway station. In: CHI 2005 Extended Abstracts on Human Factors in Computing Systems, CHI EA 2005, pp. 1637–1640. Association for Computing Machinery, New York (2005). https://doi.org/10.1145/1056808.1056985
59. Mayas, C., Steinert, T., Krömker, H., Kohlhoff, F., Hirth, M.: Challenges of operators for autonomous shuttles. In: Kromker H. (ed.) HCI in Mobility, Transport, and Automotive Systems: 5th International Conference, MobiTAS 2023, Held as Part of the 25th HCI International Conference, HCII 2023, Copenhagen, Denmark, July 23-28, 2023, Proceedings, Part I, vol. 14048 LNCS, pp. 346. Springer

Science and Business Media Deutschland GmbH (2023). https://doi.org/10.1007/978-3-031-35678-0_22
60. Mayas, C.: Towards a customizable usage requirements cycle. In: Krömker, H. (ed.) HCI in Mobility, Transport, and Automotive Systems, pp. 205–217. Springer International Publishing, Cham (2022). https://doi.org/10.1007/978-3-031-04987-3_14
61. Mirnig, A.G., Wallner, V., Gärtner, M., Meschtscherjakov, A., Tscheligi, M.: Capacity management in an automated shuttle bus: findings from a lab study. In: 12th International Conference on Automotive User Interfaces and Interactive Vehicular Applications, AutomotiveUI 2020, pp. 270–279. Association for Computing Machinery, New York (2020). https://doi.org/10.1145/3409120.3410665
62. Mirri S., Prandi C., Salomoni P., Callegati F., Campi A.: On combining crowdsourcing, sensing and open data for an accessible smart city. In: 2014 Eighth International Conference on Next Generation Mobile Apps, Services and Technologies, pp. 294–299 (2014). https://doi.org/10.1109/NGMAST.2014.59
63. Mo, B., Zhao, Z., Koutsopoulos, H., Zhao, J.: Individual mobility prediction in mass transit systems using smart card data: An interpretable activity-based hidden markov approach. IEEE Trans. Intell. Transp. Syst. **23**(8), 12014–12026 (2022). https://doi.org/10.1109/TITS.2021.3109428
64. Moore, K.R.: Public engagement in environmental impact studies: a case study of professional communication in transportation planning. IEEE Trans. Prof. Commun. **59**(3), 245–260 (2016). https://doi.org/10.1109/TPC.2016.2583278
65. Mouloudi, A., Morizet-Mahoudeaux, P.: Design process of interactive information systems. Intell. Dec. Technol. **1**(3), 127–138 (2007). https://doi.org/10.3233/IDT-2007-1303
66. Muhammad, F., Suzianti, A., Ardi, R.: Redesign of commuter line train ticket vending machine with user-centered design approach. In: Proceedings of the 3rd International Conference on Communication and Information Processing, ICCIP 2017, pp. 134–139. Association for Computing Machinery, New York (2017). https://doi.org/10.1145/3162957.3162993
67. Müller, S., Kamieth, F., Braun, A., Dutz, T., Klein, P.: User requirements for navigation assistance in public transit for elderly people. In: Proceedings of the 6th International Conference on PErvasive Technologies Related to Assistive Environments, PETRA 2013. Association for Computing Machinery, New York (2013). https://doi.org/10.1145/2504335.2504394
68. Napper, R., Elliott, P.: Design of a new route bus for Australia - investigating the value, contradictions and cost implications of vehicle improvements. In: Australasian Transport Research Forum, ATRF - Proceedings Australasian Transport Research Forum (2013)
69. Naumann, A., Grippenkoven, J., Giesemann, S., Stein, J., Dietsch, S.: Rail human factors-human-centred design for railway systems. In: IFAC Proceedings Vol. (IFAC-PapersOnline), vol. 12, pp. 330–332. IFAC Secretariat, Las Vegas, NV (2013). https://doi.org/10.3182/20130811-5-US-2037.00095, issue: PART 1 Journal Abbreviation: IFAC Proc. Vol. (IFAC-PapersOnline)
70. Oliveira, L., Birrell, S., Cain, R.: How technology can impact customer-facing train crew experiences. Ergonomics **63**(9), 1101–1115 (2020). https://doi.org/10.1080/00140139.2020.1772377
71. Oliveira, L., Fox, C., Birrell, S., Cain, R.: Analysing passengers' behaviours when boarding trains to improve rail infrastructure and technology. Robot. Comput.-Integrat. Manufact. **57**, 282–291 (2019). https://doi.org/10.1016/j.rcim.2018.12.008

72. Page, M.J., et al.: The PRISMA 2020 statement: an updated guideline for reporting systematic reviews. Systems Control Found. Appl. **10**(1), 89 (2021). https://doi.org/10.1186/s13643-021-01626-4
73. Palacin, R.: Design-led passenger environment and passenger experience. In: ACM International Conference Proceeding Series, Association for Computing Machinery (2005)
74. Panic, N., Leoncini, E., de Belvis, G., Ricciardi, W., Boccia, S.: Evaluation of the endorsement of the preferred reporting items for systematic reviews and meta-analysis (PRISMA) statement on the quality of published systematic review and meta-analyses. PLoS ONE **8**(12), e83138 (2013). https://doi.org/10.1371/journal.pone.0083138
75. Park, S.-H., Oh, S.-C. , Yeo, M.-W. : The ergonomic design that considers the user interfaces in the railroad. In: 2006 SICE-ICASE International Joint Conference, pp. 3916–3919 (2006). https://doi.org/10.1109/SICE.2006.314903
76. Patrício, L., Sangiorgi, D., Mahr, D., Čaić, M., Kalantari, S., Sundar, S.: Leveraging service design for healthcare transformation: Toward people-centered, integrated, and technology-enabled healthcare systems. J. Serv. Manag. **31**(5), 889–909 (2020). https://doi.org/10.1108/JOSM-11-2019-0332
77. Pichlmair, M., et al.: Pen-Pen: a wellbeing design to help commuters rest and relax. In: Proceedings of the Workshop on Human-Habitat for Health (H3): Human-Habitat Multimodal Interaction for Promoting Health and Well-Being in the Internet of Things Era, H3 2018. Association for Computing Machinery, New York (2018). https://doi.org/10.1145/3279963.3279966
78. Riener, A., et al.: Improving the ux for users of automated shuttle buses in public transport: investigating aspects of exterior communication and interior design. Multimodal Technol. Interact. **5**(10) (2021). https://doi.org/10.3390/mti5100061
79. Rui, A., Plewe, D., Röcker, C.: Themed passenger carriages: promoting commuters' happiness on rapid transit systems through ambient and aesthetic intelligence. Procedia Manufact. **3**, 2103–2109 (2015). https://doi.org/10.1016/j.promfg.2015.07.348
80. Rämänen, J., Riihiaho, S., Erkkilä, M., Seppälä, A.: User studies on mobile ticketing. In: Marcus, A. (ed.) DUXU 2011. LNCS, vol. 6769, pp. 630–639. Springer, Heidelberg (2011). https://doi.org/10.1007/978-3-642-21675-6_72
81. Schneider, A., Schmidt, M., Vollenwyder, B., Siegenthaler, E.: Design for distinct user groups: a case study of user-centered methods. In: Proceedings of International BCS Human-Computer Interaction Conference, HCI, vol. 2016-July. BCS Learning and Development Ltd. (2016). https://doi.org/10.14236/ewic/hci2016.92
82. Schuß, M., Manger, C., Löcken, A., Riener, A.: You'll never ride alone: Insights into women's security needs in shared automated vehicles. In: Proceedings of the 14th International Conference on Automotive User Interfaces and Interactive Vehicular Applications, AutomotiveUI 2022, pp. 13–23. Association for Computing Machinery, New York (2022). https://doi.org/10.1145/3543174.3546848
83. Schäfer, C., Zinke, R., Künzer, L., Hofinger, G., Koch, R.: Applying persona method for describing users of escape routes. Transp. Res. Procedia. **2**, 636–641 (2014). https://doi.org/10.1016/j.trpro.2014.09.106
84. Sen, R., T. Tran, S. Khaleghian, P. Pugliese, M. Sartipi, H. Neema, A. Dubey: BTE-Sim: Fast simulation environment for public transportation. In: 2022 IEEE International Conference on Big Data (Big Data), pp. 2886–2894 (2022). https://doi.org/10.1109/BigData55660.2022.10020973

85. Shen Y., Deng G.: Examination of the influential factors for intention to use shared autonomous vehicles based on discrete choice models. In: 2022 IEEE 7th International Conference on Intelligent Transportation Engineering (ICITE), pp. 308–313 (2022). https://doi.org/10.1109/ICITE56321.2022.10101459
86. Shi, Y., Taib, R., Choi, E., Chen, F.: Multimodal human-computer interfaces for incident handling in metropolitan transport management centre. In: 2006 IEEE Intelligent Transportation Systems Conference, pp. 554–559 (2006). https://doi.org/10.1109/ITSC.2006.1706799
87. Shlayan, N., Kurkcu, A., Ozbay, K.: Exploring pedestrian Bluetooth and WiFi detection at public transportation terminals. In: 2016 IEEE 19th International Conference on Intelligent Transportation Systems (ITSC), pp. 229–234 (2016). https://doi.org/10.1109/ITSC.2016.7795559
88. Siebenhandl, K., Schreder, G., Smuc, M., Mayr, E., Nagl, M.: A user-centered design approach to self-service ticket vending machines. IEEE Trans. Prof. Commun. **56**(2), 138–159 (2013). https://doi.org/10.1109/TPC.2013.2257213
89. Siricharoen, W.V.: Experiencing user-centered design (UCD) practice (case study: interactive route navigation map of Bangkok underground and sky train). In: Forbrig, P., Paternó, F., Mark Pejtersen, A. (eds.) HCIS 2010. IAICT, vol. 332, pp. 70–79. Springer, Heidelberg (2010). https://doi.org/10.1007/978-3-642-15231-3_8
90. Siricharoen, W.: Operating human computer interaction approach in design phases of software developing process. J. Converg. Inf. Technol. **7**(11), 422–431 (2012). https://doi.org/10.4156/jcit.vol7.issue11.53
91. Stankov, U., Gretzel, U.: Tourism 4.0 technologies and tourist experiences: a human-centered design perspective. Inform. Technol. Tourism **22**, 477 – 488 (2020)
92. Stein, M., Meurer, J., Boden, A., Wulf, V.: Mobility in later life: appropriation of an integrated transportation platform. In: Proceedings of the 2017 CHI Conference on Human Factors in Computing Systems, CHI 2017, pp. 5716–5729. Association for Computing Machinery, New York (2017). https://doi.org/10.1145/3025453.3025672
93. Stjernborg, V., Mattisson, O.: The role of public transport in society-a case study of general policy documents in Sweden. Sustainability **8**(11) (2016). https://doi.org/10.3390/su8111120
94. Stockinger, C., König, C.: User-centered development of a support-system for visually handicapped people in the context of public transportation. In: Bagnara, S., Tartaglia, R., Albolino, S., Alexander, T., Fujita, Y. (eds.) IEA 2018. AISC, vol. 824, pp. 1473–1482. Springer, Cham (2019). https://doi.org/10.1007/978-3-319-96071-5_150
95. Stoop, J., Baggen, J., Vrancken, J., Vleugel, J., Beukenkamp, W.: A diabolic dilemma: Towards fully automated train control or a human centred design? In: IFAC Proc. Vol. (IFAC-PapersOnline). vol. 42, pp. 251–256. IFAC Secretariat (2009). https://doi.org/10.3182/20090902-3-US-2007.0001, issue: 15 Journal Abbreviation: IFAC Proc. Vol. (IFAC-PapersOnline)
96. Stoop, J., Baggen, J., Vleugel, J., Vrancken, J.: ERTMS, deals on wheels? An inquiry into a major railway project. In: Saf., Reliability Risk Anal.: Theory, Methods Appl. - Proc. Jt.ESREL SRA- Eur. Conf, vol. 2, Valencia, pp. 1519–1524 (2009)
97. Sya'Bana, Y., Sanjaya, K.: The applicability of sustainable design values on electric bike sharing concept in Indonesia. In: Pramana R.I., Atmaja T.D., Putrasari

Y., Muqorobin A., Sadono A.P., Sudirja S., Mirdanies M., Martides E. (eds.) Proceeding - International Conference on Sustainable Energy Engineering and Application, Innovative Technology Toward Energy Resilience, ICSEEA, pp. 125–130. Institute of Electrical and Electronics Engineers Inc. (2019). https://doi.org/10.1109/ICSEEA47812.2019.8938654

98. Tan, H., Sun, J., Wenjia, W., Zhu, C.: User experience & usability of driving: a bibliometric analysis of 2000–2019. Inter. J. Hum.-Comput. Interact. **37**(4), 297–307 (2021). https://doi.org/10.1080/10447318.2020.1860516
99. Teo, Z.T., et al.: SecureRails: towards an open simulation platform for analyzing cyber-physical attacks in railways. In: 2016 IEEE Region 10 Conference (TENCON), pp. 95–98 (2017). https://doi.org/10.1109/TENCON.2016.7847966
100. Toledo, J., Salgado, A.C., Gozum, C.C., Custodio, B.: Usability assessment of a philippine commuting navigation application. In: Ahram, T., Falcão, C. (eds.) AHFE 2020. AISC, vol. 1217, pp. 286–293. Springer, Cham (2020). https://doi.org/10.1007/978-3-030-51828-8_38
101. Tucho, G.T.: A review on the socio-economic impacts of informal transportation and its complementarity to address equity and achieve sustainable development goals. J. Eng. Appli. Sci. **69**(1) (2022). https://doi.org/10.1186/s44147-022-00074-8
102. Vallet, F., Khouadjia, M., Amrani, A., Pouzet, J.: Designing a data visualisation and analysis tool for supporting decision-making with public transportation network. Proc. Des. Soc. vol. 1, pp. 1093–1102 (2021). https://doi.org/10.1017/pds.2021.109
103. Vargas-Acosta, R.A., Becerra, D.L., Gurbuz, O., Villanueva-Rosales, N., Nunez-Mchiri, G.G., Cheu, R.L.: Smart mobility for seniors through the urban connector. In: 2019 IEEE International Smart Cities Conference (ISC2), pp. 241–246 (2019). https://doi.org/10.1109/ISC246665.2019.9071732
104. Vernier M., Redmill K., Ozguner U., Kurt A., GuvencB. A.: OSU SMOOTH in a smart city. In: 2016 1st International Workshop on Science of Smart City Operations and Platforms Engineering (SCOPE) in partnership with Global City Teams Challenge (GCTC) (SCOPE - GCTC), pp. 1–6 (2016). https://doi.org/10.1109/SCOPE.2016.7515057
105. Wagner, I., Basile, M., Ehrenstrasser, L., Maquil, V., Terrin, J.J., Wagner, M.: Supporting community engagement in the city: Urban planning in the MR-tent. In: Proceedings of the Fourth International Conference on Communities and Technologies, C&T 2009, pp. 185–194. Association for Computing Machinery, New York (2009). https://doi.org/10.1145/1556460.1556488
106. Wang, J., Fang, W., Guo, B., Niu, K.: Function allocation design of subway automatic train supervision system's alarm unit. In: 2019 IEEE International Conference on Industrial Engineering and Engineering Management (IEEM), pp. 45–49 (2019). https://doi.org/10.1109/IEEM44572.2019.8978768
107. Wang, Z., dos Santos, M., Cheung, V.: An eye-tracking study examining information search in transit maps using China's high-speed railway map as a case study. Inform. Design J. **26**(1), 53–72 (2021). https://doi.org/10.1075/idj.20011.wan
108. Wilkowska, W., Arning, K., Ziefle, M.: Participatory design in the development of a smart pedestrian mobility device for urban spaces. In: Marcus, A., Wang, W. (eds.) DUXU 2017. LNCS, vol. 10290, pp. 753–772. Springer, Cham (2017). https://doi.org/10.1007/978-3-319-58640-3_54
109. Wintersberger, P., Frison, A.K., Riener, A.: Man vs. Machine: comparing a fully automated bus shuttle with a manually driven group taxi in a field study. In:

Adjunct Proceedings of the 10th International Conference on Automotive User Interfaces and Interactive Vehicular Applications, AutomotiveUI 2018, pp. 215–220. Association for Computing Machinery, New York (2018). https://doi.org/10.1145/3239092.3265969

110. Wu, J., Johnson, S., Hesseldahl, K., Quinlan, D., Zileli, S., Harrow, P.D.: Defining ritualistic driver and passenger behaviour to inform in-vehicle experiences. In: Adjunct Proceedings of the 10th International Conference on Automotive User Interfaces and Interactive Vehicular Applications, AutomotiveUI 2018, pp. 72–76. Association for Computing Machinery, New York (2018). https://doi.org/10.1145/3239092.3265944
111. Wu, D., et al.: Enabling efficient offline mobile access to online social media on urban underground metro systems. IEEE Trans. Intell. Transp. Syst. **21**(7), 2750–2764 (2020). https://doi.org/10.1109/TITS.2019.2911624
112. Yamada-Kawai, K., Hirasawa, N., Fukada, H., Ohtsu, S.: Implication of envisioning citizen-centered administrative services. In: Proceedings of SICE Annual Conference 2010, pp. 1437–1440 (2010)
113. Yang, L., Zhu, Y., Chatzimichailidou, M., Liu, X.: Assessing human emotional responses to the design of public spaces around subway stations: a human factors research. Urban Des. Inter. **28**(4), 285–303 (2023). https://doi.org/10.1057/s41289-023-00219-y
114. Yang H., Chiang C. -F., Arbee C. L.P.: Discovering high demanding bus routes using farecard data. In: 2019 IEEE International Conference on Big Data (Big Data), pp. 3832–3837 (2019). https://doi.org/10.1109/BigData47090.2019.9006288
115. Yoo, D., Zimmerman, J., Hirsch, T.: Probing bus stop for insights on transit co-design. In: Proceedings of the SIGCHI Conference on Human Factors in Computing Systems, CHI 2013, pp. 409–418. Association for Computing Machinery, New York (2013). https://doi.org/10.1145/2470654.2470714
116. Zahler T.: Characteristics of human behaviour in safety-critical systems by the example of european railway control centres. In: 2008 3rd IET International Conference on System Safety, pp. 1–6 (2008). https://doi.org/10.1049/cp:20080708
117. Zhang, T., Chen, J., Yang, S., Song, P., Qi, H.: The application of non-newtonian fluid speed reducer in human-oriented traffic concept. In: 2019 4th International Conference on Electromechanical Control Technology and Transportation (ICECTT), pp. 357–360 (2019). https://doi.org/10.1109/ICECTT.2019.00087
118. Zhao, T., Siu, K.: The boundaries of public space: a case study of Hong Kong's Mass Transit Railway. Int. J. Des. **8**(2), 43–60 (2014)
119. Zhao, W., Jiang, H., Tang, K., Pei, W., Wu, Y., Qayoom, A.: Knotted-line: a Visual explorer for uncertainty in transportation system. J. Comput. Lang. **53**, 1–8 (2019). https://doi.org/10.1016/j.cola.2019.01.001
120. Zhao, C., Dai, X., Lv, Y., Niu, J., Lin, Y.: Decentralized autonomous operations and organizations in TransVerse: federated intelligence for smart mobility. IEEE Trans. Syst. Man Cybernet. Syst. **53**(4), 2062–2072 (2023). https://doi.org/10.1109/TSMC.2022.3228914
121. Zimmerman, J., et al.: Field trial of Tiramisu: crowd-sourcing bus arrival times to spur co-design. In: Proceedings of the SIGCHI Conference on Human Factors in Computing Systems, CHI 2011, pp. 1677–1686. Association for Computing Machinery, New York (2011). https://doi.org/10.1145/1978942.1979187

Pedestrian Interaction with Automated Driving Systems: Acceptance Model and Design of External Communication Interface

Viktoria Marcus, Joseph Muldoon, and Sanaz Motamedi(✉)

University of Florida, Gainesville, FL 32611, USA
{viktoria.marcus,smotamedi}@ufl.edu

Abstract. In 2021, almost 70,000 pedestrians were injured or killed in traffic accidents in the United States [1]. Level-5 Automated Driving Systems (ADSs) have the potential to create safer roads by eliminating human errors [2] system. While the development and deployment of level-5 ADSs has been improved, the interactions between pedestrians and ADSs are not fully understood, despite pedestrians being the most vulnerable road users. This study investigated the factors that affect pedestrians' acceptance of level-5 ADSs and design features for safe and efficient external human-machine interfaces (eHMIs) to facilitate communication. A survey was conducted with 37 participants to investigate the impact of pedestrians' background, behaviors, and personal innovativeness on ADS acceptance. A follow-up lab study was performed with 70 participants to determine effective eHMI design features. It was found that there was no effect of background information on the acceptance factors or behavioral intention to cross in front of level-5 ADSs, though pedestrian behaviors and personal innovativeness had a significant effect. In the eHMI lab study, both visual and auditory features were used to create eHMIs including external speedometers, audio cues giving advice to pedestrians, and a method of indicating the driving system was level-5 ADSs. This study gives recommendations about the effect of pedestrians' background, behaviors, and personal innovativeness on eHMI acceptance and intention to cross the street in front of level-5 ADSs as well as several key visual and auditory features pedestrians included in eHMIs.

Keyword: pedestrian interaction with level-5 ADS · eHMI · safety · trust

1 Introduction

In 2021, almost 70,000 pedestrians were injured or killed in traffic crashes in the United States, an increase from the previous years [1]. This amounted to 20 pedestrian deaths per day, and 142 per week, a problem that the National Highway Traffic Safety Administration has called "a national crisis of traffic deaths" [1, 3]. Many of these damages, injuries, and deaths are caused by human error, whether it is the fault of the pedestrian, driver, or both; most pedestrian fatalities occurred in complex traffic scenarios–in busy urban areas, at intersections, and in darkness [1, 4, 5]. Level-5 Automated Driving Systems

H. Krömker (Ed.): HCII 2024, LNCS 14733, pp. 63–82, 2024.
https://doi.org/10.1007/978-3-031-60480-5_4

(ADSs), defined as automated driving systems with the ability to perform "sustained and unconditional" operation of a vehicle "under all road conditions in which a conventional vehicle can be reasonably operated by a … human driver," have the potential to reduce the risks caused by driver error and create safer road conditions for all users, especially pedestrians, by eliminating the need for a human operator [2]. However, this is only possible if pedestrians can easily accept and understand the intentions of level-5 ADSs. Currently, the interactions between pedestrians and level-5 ADSs, as well as pedestrians' attitudes towards level-5 ADSs, are not well understood.

Pedestrians often face complex and risky traffic scenarios, such as crossing the road in front of a vehicle at uncontrolled intersections. In such events, pedestrians need to make a safe decision by gathering information about the vehicle they are about to step in front of. When pedestrians cross in front of human-driven vehicles (HDVs) they often rely on verbal and/or non-verbal communication with the driver, including eye-contact and hand gestures, to understand what the driver's intentions are [6–14]. But when there is no driver, this kind of communication is unavailable, yet the pedestrian still needs enough information to be able to make a safe crossing decision. Without adequate information about the vehicle's intentions, the pedestrian may make an unsafe choice, or the vehicle and pedestrian may be caught in a deadlock, where neither move [12].

To facilitate effective communication between level-5 ADSs and pedestrians, external human-machine interfaces, eHMIs, have been proposed [8, 9, 11, 14–20]. eHMIs often use different modalities for communicating to pedestrians, including various visual and aural features, ranging from inset eyes to LED text grilles to musical chimes; while many different eHMI designs have been suggested, no standardized design has been widely tested or accepted. Previous studies have shown that the inclusion of any kind of eHMI, whether it uses visual or aural features, gives instructions to pedestrians, or communicates the level-5 ADS' intentions, have helped pedestrians clearly and safely interact with level-5 ADSs [8, 9, 13–20]. Many features have been studied in eHMI design, though few, if any, studies have directly asked pedestrians what features they would most prefer, and what information they find critical to making safe road-crossing decisions. A standardized eHMI design is key for the acceptance of level-5 ADSs, as clear, consistent communication with them will help pedestrians gather enough information about their intentions to make safe crossing decisions.

In addition, it is key to understand the factors that influence pedestrians' decisions to cross in front of a vehicle, as many conditions may affect this decision, and the most efficient eHMI design must be able to take these factors into consideration. When pedestrians consider crossing in front of any kind of vehicle, they need to know who has the right of way (ROW), if traffic is approaching and if it is slowing down or speeding up, what the road and environmental conditions are, if other road users are present, what infrastructure is around them, and other information that will affect their ability to safely cross [6–8, 11–16, 18, 20–25]. But specifically with level-5 ADSs, pedestrians need to know if the level-5 ADS has detected them and if it will yield; studies have shown that pedestrians are prone to over-trusting level-5 ADSs, assuming that all automated driving systems will follow a more conservative driving style and always give them the ROW, leading to pedestrians taking more risks around level-5 ADSs [12, 16, 22]. Few studies have investigated what factors affect pedestrians' intentions to cross in front of level-5

ADSs. There is a lack of user-centric studies to identify efficient eHMI designs and develop a theoretical framework for pedestrian behavioral intention to cross the road in front of level-5 ADSs.

This study seeks to address the gaps in understanding of which factors influence pedestrians' crossing decisions and eHMI preferences, by developing a theoretical framework describing pedestrians' intentions to cross the road in front of level-5 ADSs and conducting a lab study and using human center design approach to generate eHMI designs to understand which features and modalities they most preferred.

2 Literature Review

To develop a foundational understanding of pedestrian behavior and acceptance of level-5 ADSs, pedestrian interactions with level-5 ADSs, and eHMI design considerations, a literature review was first performed. The papers examined in this study related to theories of behavior and technology acceptance, level-5 ADS acceptance, pedestrian interaction with HDVs and level-5 ADSs, and the impact of eHMIs on pedestrian interaction with level-5 ADSs.

2.1 Theories of Behavior and Level-5 ADS Acceptance

There are many different behavioral models that establish relationships between external factors, intention to perform an action, and the actual performance of the action; among these, the Theory of Planned Behavior (TPB) [26], Technology Acceptance Model (TAM) [27], Unified Theory of Acceptance and Use of Technology (UTAUT) [28], and several others, have been widely used, verified, and extended. Kaye et al. [15] applied the TPB, TAM, and UTAUT models through surveys The study found that TPB and UTAUT explained the most variance in intentions to cross the road in front of a level-5 ADS; it suggests that TPB can be used as a base to explore pedestrians' acceptance of level-5 ADSs.

Another aspect of acceptance, beside TPB factors, is to understand what risky behaviors pedestrians exhibit; these behaviors can be described through five categories: violation, error, lapse, aggressive behavior, and positive behavior [25]. These five categories are included in the Pedestrian Behavior Questionnaire (PBQ), developed by [25], with the goal of being able to identify how pedestrians interact with other road-users, infrastructure, and other pedestrians. To apply the PBQ to behavioral models describing pedestrians' receptivity of level-5 ADSs, the study Deb et al., [25] developed and validated a pedestrian receptivity questionnaire including the factors of attitude, social norm, trust, compatibility, system effectiveness, personal innovativeness, and pedestrian behaviors via the PBQ.

2.2 Pedestrian Interaction with HDVs, Level-5 ADSs, and eHMIs

Understanding the ways the pedestrians interact with conventional vehicles and level-5 ADSs, as well as understanding the effect of eHMIs, can be done by reviewing studies that used surveys or practical trials to determine pedestrians' real-life interaction with various types of vehicles. These studies show the differences pedestrians exhibit when interacting with HDVs versus level-5 ADSs [10, 22–24]. While the studies indicate that pedestrians do tend to trust level-5 ADSs more than HDVs, and [10] found that pedestrians crossed in front of a simulated level-5 ADS in a manner similar to an HDV, it is still vital to understand how to facilitate effective communication between pedestrians and level-5 autonomous driving systems.

To analyze and understand the impact of an eHMI in facilitating communication between pedestrians and level-5 ADSs, many studies have utilized virtual reality experiments, practical trials, videos, or surveys to measure how pedestrians respond to various eHMIs [7–9, 13, 16, 18, 20]. Replacing interactions with the human driver, eHMIs are meant to assist pedestrians in making safe road-crossing decisions, by presenting information about the intention of the level-5 ADS or telling the pedestrian what to do through visual and aural features. Studies investigating eHMI communication often analyze both how pedestrians feel when interacting with ADSs (utilizing behavioral acceptance factors, such as safety or trust) and how eHMIs were able to facilitate more efficient interactions [7–9, 11–14, 16–20].

These studies have found that eHMIs utilizing multiple communication modalities, including both visual and auditory features are preferred by pedestrians, especially when they use symbols and colors found in existing traffic infrastructure, such as pedestrian silhouettes or red blinking hands, as well as text displays, which were often rated highly despite them being language dependent [8, 13, 16].Studies also found that pedestrians relied on implicit communication from the level-5 ADS, such as its speed, before they used the eHMI; pedestrians used the eHMI to verify the level-5 ADSs' intention, which they first determine from the vehicle's movement [7, 11, 13, 17, 18]. Overall, it was found that eHMIs have a positive impact on pedestrian interaction with level-5 ADSs [9, 12, 13, 16].

These studies analyzed how pedestrians interact with ADSs and various eHMIs, but a common problem identified by the authors is pedestrians' lack of experience with level-5 ADS, and the fact that there is no standardized eHMI design, as well as the challenges with designing eHMIs that are accessible to the widest possible population, including pedestrians from different countries, speaking different languages, or are visually or aurally impaired [8, 9, 11–22, 24, 25]. Additionally, many studies found that implicit communication through movement and speed was often more important than explicit communication through eHMIs, and that eHMIs had to correspond to the movement of the level-5 ADSs or add additional information in order to be useful [6–8, 11, 14–17, 19]. Overall, many of the studies did show that eHMIs increased pedestrians' perceptions of safety when interacting with level-5 ADSs.

2.3 Further eHMI Design Considerations

To try to determine the most effective eHMI design, several studies have compared eHMI modalities, iterated on specific designs, tried to find the most effective location on a level-5 ADS to include an eHMI, and even studied the effect of infrastructure and environment on eHMI use.

These studies found that eHMIs should be placed in multiple locations on a level-5 ADS, to ensure that information could be viewed by pedestrians viewing the vehicle at any angle [19]. And, in especially complex traffic scenarios, [20] found that eHMIs were especially helpful to pedestrians in making safe road-crossing decisions. When investigating the communication style of text-based eHMIs [14] found that polite wording led to more compliance and positively affected trust and acceptance of the level-5 ADS.

While these studies [7–9, 13, 14, 16–20] explored different designs, many face common limitations. Most often, these studies do not include large or diverse sample groups, with many having samples composed of pedestrians without any disabilities (such as visual or auditory impairment, which may change how pedestrians interact with eHMIs) [8–10, 15, 16, 22–25]. This causes many of these studies to not be widely generalizable. In addition, many studies provide simplified conditions for interaction, such as having only one level-5 ADS interacting with one pedestrian, or only considering a single lane road [8, 11, 18, 20]. Real traffic scenarios are much more complex and dynamic, which may make interactions with eHMIs much different. Additionally, many pedestrians do not have much experience interacting with level-5 ADSs, meaning that their interactions may change over time as they become more and more mainstream.

The mentioned studies provide critical context for pedestrian interaction and acceptance of level-5 ADSs and various eHMIs, but none of the studies used human centric design methods to identify the pedestrians' opinion about what eHMI features and what information is required for to support them in deciding to cross the road in risky situation (e.g. crossing at uncontrolled intersection). Many studies find that eHMIs are overall beneficial, and some ask pedestrians to rate existing eHMI designs, but the results concerning which eHMI design would be the most efficient are inconclusive, as most studies find no statistically significant differences between various eHMIs' effects on crossing behavior, despite some being ranked higher than others for preference [6, 9, 13, 14, 16–19]. To address these gaps, this study explores the factors important to pedestrians' acceptance of level-5 ADSs, pedestrian behavior, personal innovativeness, background, and what information and features pedestrians most needed in an eHMI when interacting with a level-5 ADS in risky situation (e.g. uncontrolled intersection).

3 Theoretical Framework

The Theory of Planned Behavior, TPB, has been shown to be useful in describing pedestrians' intentions to cross the road in front of a level-5 ADS, and has been extended with other factors to create models that describe pedestrians' behaviors [15, 22, 25, 26]. A goal of this study is to create a theoretical framework describing what factors impact pedestrians' acceptance of level-5 ADSs by extending the TPB with additional factors. The model is shown in Fig. 1.

The base TPB model includes the factors of perceived behavioral control, or PBC, social norm, and attitudes, which are predictors of behavioral intention and actual behavior [26]. Following the results from literature review, to extend the model to address the objectives of this study, the TPB was extended with the external factors of background information, personal innovativeness, general pedestrian behaviors, and external acceptance factors [15, 21, 22, 24].

Background information factors included gender and previous knowledge of level-5 ADSs. Previous studies have had mixed results regarding the impact of age, gender, and previous experience [15, 16, 24].

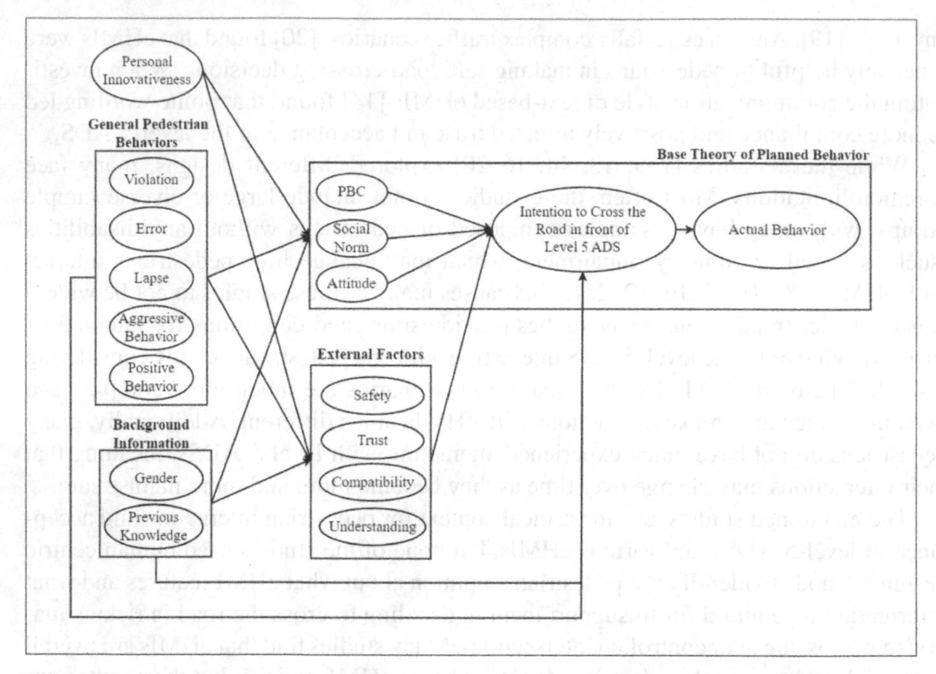

Fig. 1. Theoretical framework based on TPB, extended with personal innovativeness, general pedestrian behaviors, background information, and external factors.

Personal innovativeness was also included in this model, as previous studies have shown that pedestrians with more inclinations towards adopting new technology have more acceptance of level-5 ADSs and more positive attitudes towards them [15, 21].

The general pedestrian behaviors factors included violation, lapse, error, aggressive behaviors, and positive behaviors, based on the results from the study by [21]. Pedestrian behaviors were included in this study because pedestrians often assume that level-5 ADSs will follow a more conservative driving style and always yield to them, leading to pedestrians engaging in more risky behaviors around them [7, 22]. Poor traffic decisions made by pedestrians indicate a lack of control, situational awareness, and predictability, the last of which is important for level-5 ADSs to be able to understand pedestrians' intentions [21].

The acceptance factors used in the model were safety, trust, compatibility, and understanding [15, 21]. The definitions of all the factors used in the model are shown in Table 1.

Table 1. Factors in theoretical framework.

Factor	Definition	Reference
Background Information		
Previous knowledge of level-5 ADSs	Pedestrians' existing understanding, experience with, or interest in level-5 ADSs	[16]
General Pedestrian Behaviors		
Violation	"Deliberate deviation from social rules without intention to cause injury or damage."	[21]
Error	"Deficiency in knowledge of traffic rules and/or in the inferential processes involved in making a decision."	[21]
Lapse	"Unintentional deviation from practices related to a lack of concentration on the task; forgetfulness."	[21]
Aggressive Behavior	"A tendency to misinterpret other road users' behavior resulting in the intention to annoy or endanger."	[21]
Positive Behavior	"Behavior that seeks to avoid violation or error and/or seeks to ensure traffic rule compliance."	[21]
Personal Innovativeness		
Personal Innovativeness	Willingness to try something new or adopt a new technology	[15, 21]
Theory of Planned Behavior		
Perceived Behavioral Control	A pedestrian's feeling of their level of control in a road-crossing scenario with a level-5 ADS, including how easy it would be to cross in front of the level-5 ADS, and if the decision were mostly up to them	[15, 21]
Social Norm	How strongly a pedestrian values the opinions of those important to them when considering the road in front of a level-5 ADS, as well as if their decision to cross would be influenced by their loved ones	[15, 21]

(*continued*)

Table 1. (*continued*)

Factor	Definition	Reference
Attitude	A pedestrian's feeling towards level-5 ADSs in general, including if they think level-5 ADSs are a good idea, would enhance the overall transportation system, and/or make roads safer for all road users	[15, 21]
Intention to Cross	A pedestrian's future willingness, ability, and feeling towards crossing the road in front of a level-5 ADS	[15]
Acceptance Factors		
Safety	How risky a pedestrian would feel crossing in front of a level-5 ADS, their general feelings of safety around a level-5 ADS, and if they would feel more or less safe crossing the road in front of a level-5 ADS or an HDV	[15]
Trust	The level of comfort a pedestrian would feel when considering themself or a loved one crossing the street in front of a level-5 ADS, and their general level of trust in level-5 ADSs	[21]
Compatibility	How well a pedestrian believes that level-5 ADSs would be able to integrate into the existing traffic infrastructure	[15, 21]
Understanding	How well a pedestrian believes they would be able to interact with an understand the intentions of a level-5 ADS, and if they believe level-5 ADSs would be able to effectively communicate and interact with other pedestrians	[15, 16, 21]

4 Methodology

This study was conducted in two parts: a survey measuring pedestrians' acceptance of level-5 ADSs through their behavioral intention to cross the road in front of a level-5 ADS, and a lab study conducted to generate eHMI designs tailored to pedestrians' preferences. The study was approved by the International Review Board, IRB.

4.1 Participants

The participants in this study were all volunteers. The sample was composed of industrial and systems engineering students attending a large, public university in the USA. All recruited volunteers had been trained for about 6 h regarding human centered design and data collection through interviews and surveys. We collected 37 valid survey responses from the volunteers. All participants were between the ages of 18–24; it is for this reason that age, a commonly used factor in similar studies, was omitted. Of the 37 valid respondents, 22 were male and 15 were female. 94.6% of the participants had previous knowledge of level-5 ADSs prior to taking the survey and participating in the lab study. In the lab study, all 70 volunteers participated, split into 25 groups. It is worth noting that all volunteers who responded to the survey participated in the lab study.

4.2 Online Acceptance Survey

To measure pedestrians' intention to cross in front of level-5 ADSs, an online survey was conducted. The survey instrument was made to investigate the theoretical framework in Fig. 1. The survey was created using the Qualtrics software and contained 62 questions in the following sections: informed consent, background information, general pedestrian behaviors, personal innovativeness, and the acceptance factors.

In the informed consent section of the survey, the purpose of the study was explained and a definition of level-5 ADSs was given, prior to giving consent to participation. The background information section included items asking about students' age, gender, and previous knowledge of level-5 ADSs. The general pedestrian behaviors section followed the PBQ, or pedestrian behavior questionnaire, developed by [25]. The personal innovativeness section of the survey used the same four-item scale that [21] used. The final section of the survey asked after the acceptance factors and TPB factors, including PBC, social norm, attitude, safety, trust, compatibility, and understanding, was randomized.

4.3 EHMI Design Lab

The following day after the survey was distributed, 70 volunteers participated in a lab design study to develop eHMIs based on pedestrian needs.

During the lab study, the students were first introduced to level-5 ADSs and eHMIs, including their potential benefits and risks. Students were asked to think about what information pedestrians would need from level-5 ADSs as they were deciding whether to cross the road in front of one.

The students were also given training in user-centric design, with the three steps of understanding user needs, designing a prototype, and evaluating the design described as well as how to conduct interviews and surveys.

The students then split into groups of three and took turns each filling in the roles of interviewee, facilitator, and moderator. The interviewee student would answer prewritten questions asked by the facilitator from the perspective of a pedestrian, the facilitator would record the answer, and the moderator would keep the rest of the group on time and on task. Each student participated in each role once, and the responses to the questions were summarized.

The students responded to the following questions:

1. Describe your road crossing behavior in general.

2. Imagine that you are crossing the road while a level-5 ADS is driving towards you. What information do you need from the level-5 ADS to make a decision about crossing or not crossing the road, and why?
3. How do you want this information communicated?
4. When should this information be communicated?
5. What would make you feel safe crossing in front of a level-5 ADS?
6. What would make you trust a level-5 ADS?

   - The purpose of this activity was to give the participants an understanding of the needs of a pedestrian when they are attempting to cross the road in front of a level-5 ADS. Based on the interviews, each student would create a list of requirements for an eHMI design that was meant to facilitate efficient interactions between pedestrians and level-5 ADSs, and tailor it to the needs of the pedestrian.
   - Next, each student would create a sketch of an eHMI based on a different group member's list of requirements. The sketches were annotated and labeled to ensure all required features were met. Then, the group members either picked the best design of the group or combined all three into one eHMI design. The groups wrote justifications for why those designs were selected.
   - Then, two groups of three would merge into one group of six, and each group would present their designs to the others. The group members would then ask questions about the designs to understand why those specific features and locations were chosen, with the goal of refining and improving the designs. Lastly, the groups would submit their final eHMI designs and a summary of the user's (pedestrian's) needs and what key features they included in their eHMI designs.

5 Results

5.1 Survey Data Analysis

Effect of Background Information. To understand the effect of background information on acceptance of level-5 ADSs and intention to cross the road in front of a level-5 ADS, two-way ANOVA was conducted with gender and previous knowledge of level-5 ADSs as independent variables and the acceptance factors (safety, trust, PBC, attitude, social norm, compatibility, understanding, and intention to cross the road in front of level-5 ADSs) as dependent variables. Age was not included in the analysis, as all the participants were in the same age range, 18–24 years. The results from the two-way ANOVA conducted with gender and previous knowledge of level-5 ADSs had no effect on the acceptance factors. The detailed results including p-values for each ANOVA test are shown in Table 2.

Effect of Pedestrian Behavior. To understand the effect of general pedestrian behavior, factorial ANOVA was conducted with violation, lapse, error, aggressive behavior, and positive behavior as the independent variables, and the acceptance factors as dependent variables. The results from the factorial ANOVA showed that violation and error had a significant effect on safety ($p = 0.01$); error and positive behavior had a significant effect on trust ($p = 0.01$, $p = 0.1$); error, lapse, and positive behavior had a significant effect on PBC ($p = 0.01$); lapse had a significant effect on attitude ($p = 0.01$); error had a significant effect on understanding (p = 0.1); lapse and positive behavior had a significant effect on social norm (p = 0.01); positive behavior had a significant effect on compatibility (p = 0.001); and error and lapse had a significant effect on behavioral intention to cross in front of a level-5 ADS (p = 0.001, p = 0.01).

Error had the most influence over the acceptance factors, having a statistically significant effect on all factors but attitude, social norm, and compatibility. Aggressive behavior had no effect on any of the acceptance factors or intention to cross. The detailed results are shown in Table 3.

Effect of Personal Innovativeness. To understand the effect of personal innovativeness on acceptance and intention to cross, one-way ANOVA was conducted with personal innovativeness as the independent variable, and the acceptance factors as the dependent variables. The one-way ANOVA conducted with personal innovativeness showed that personal innovativeness had a significant effect on trust (p = 0.01), PBC (p = 0.1), attitude (p = 0.01), social norm (p = 0.01), and compatibility (p = 0.01). Personal innovativeness did not influence safety, understanding, or behavioral intention to cross. The detailed results are shown in Tables 4.

Table 2. Results of ANOVA analysis between Gender, Experience, and Acceptance Factors.

ANOVA with Gender, Experience, and Acceptance Factors								
	Safety	Trust	PBC	Attitude	Social Norm	Compatibility	Understanding	BI
	F-value	F-value	F-value	F-value	F-value	F-value	F-value	F-value
Gender	0.119	0.957	0.358	0.089	0.053	0.105	1.135	1.093
Experience	0.228	1.039	2.148	1.675	0.017	1.841	1.319	0.824

Table 3. Results of ANOVA analysis between Pedestrian Behaviors and Acceptance Factors.

ANOVA with Pedestrian Behaviors and Acceptance Factors								
	Safety	Trust	PBC	Attitude	Social Norm	Compatibility	Understanding	BI
	F-value	F-value	F-value	F-value	F-value	F-value	F-value	F-value
Violation	4.270*	0.770	0.318	0.511	0.036	0.374	1.272	1.074
Error	4.651*	5.524*	6.408**	2.333	0.141	2.074	3.036	10.109**
Lapse	0.054	1.915	6.077*	4.567**	6.140**	2.254	0.913	5.002*
Aggressive	0.033	0.683	0.096	0.359	1.316	0.347	0.112	0.463
Positive	3.098	3.475	5.888*	2.689	6.260*	7.625***	0.769	0.790

5.2 EHMI Design Lab Results

The eHMI design lab generated 25 different eHMI designs, emphasizing meeting pedestrians' needs. Prior to creating eHMI designs, the participants first identified exactly what

Table 4. Results of ANOVA analysis between Personal Innovativeness and Acceptance Factors.

ANOVA with Personal Innovativeness and Acceptance Factors								
	Safety	Trust	PBC	Attitude	Social Norm	Compatibility	Understanding	BI
	F-value	F-value	F-value	F-value	F-value	F-value	F-value	F-value
PI	0.001	6.417**	3.347	4.137*	7.136*	5.912**	2.487	0.661

'*' in the above tables show the degrees of freedom and p-value results from ANOVA. '***' means the p-value is less than 0.01, '**' means the p-value is less than 0.02 significant and '*' means the p-value is less than 0.05.

needs pedestrians had when crossing the road in front of a level-5 ADS. Participants were asked to consider what information pedestrians needed to make a safe crossing decision, as well as what would make them feel safe around a level-5 ADS and what would increase their trust in.

Many participants agreed that they needed a clear indication that the level-5 ADS had identified them, that the level-5 ADS was slowing down to a stop, and required clear signals telling the pedestrians when it was safe to cross. Participants also wanted to know information about the state of the level-5 ADS and if its intentions were changing, especially knowing the speed of the vehicle and if it were changing (speeding up or slowing down); one group stated that they wanted "evidence the [level-5 ADS] sees [them] crossing and doesn't move from the period [they] are on the side of the road until [they] are on the other side of the road." Other groups reported that "if the vehicle reduces its speed and generally comes to a stop, it shows [them] that it is safe to cross," as "coming to a full stop feels safe."

Regarding feelings of safety and trust, participants reiterated the importance of "knowing that the [level-5 ADS] sees" the pedestrian and "would feel safe crossing in front of [a level-5 ADS] if information about its speed and movement were communicated" and they were given "very clear directions when it is safe to cross or not." The students clearly stated that they would "feel safe if ... the [level-5 ADS] acknowledged [their] presence and showed signs of stopping." Another group restated the importance of knowing the level-5 ADSs has seen them, saying, "if the vehicle shows that it is aware of the pedestrian, it would be safe to cross."

The students also stated more broad desires to know that level-5 ADSs had gone through comprehensive testing. One group specified that extensive "design testing and transparent communication of the testing results would make [them] trust [a level-5 ADS] since [they would be] properly informed of its safety features." Participants mentioned that "lots of testing," "good public opinion," and "years of successful field testing" in "areas that have little and heavy pedestrian traffic" would help improve their trust in level-5 ADSs. Students also mentioned that real-world "experience" with level-5 ADSs would increase their levels of trust, as they would get to learn how the system worked; others said that "understanding how the autonomous system works could give [them] more confidence in its abilities."

Once these needs were identified, the participants generated 25 eHMI designs. The features were categorized according to the guidelines used in Wang et al. [11]. These categories are: 1) intent of the ADS, 2) advice to pedestrians of what to do, 3) the ADS's awareness of pedestrians, 4) combination of intent and advice, and 5) ADS movement (8).

Within these categories, data was split up into two additional groups: one relating to audio features (Fig. 2) and the other with visual features (Fig. 3). For visual features, the location on the level-5 ADS of each feature was also noted. Participants were asked to justify the features they chose to include in their designs. For many, enhancing safety and trust was at the forefront of their considerations for both visual and auditory features.

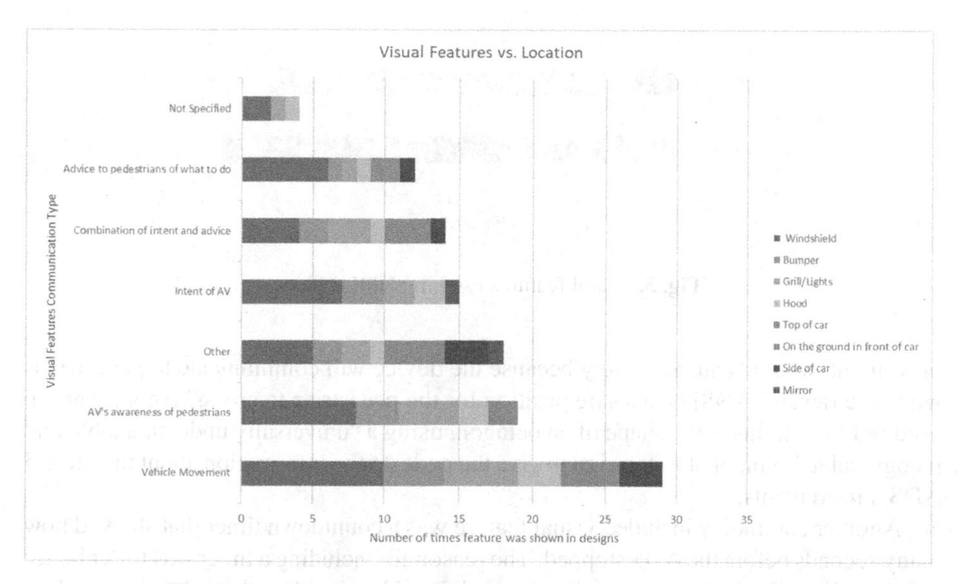

Fig. 2. Visual features and their locations used in eHMI designs.

eHMI design elements were grouped based on how they communicated their information. For example, a sign with words telling the pedestrian to go would fall under "advice to pedestrians of what to do."

The most common specific visual features participants included in their eHMIs were external speedometers, LED lights, countdown timers, and Level-5 ADS labels. The most common specific auditory features included in eHMI designs were beeping, jingles, and spoken words.

Speed indicators, including external speedometers, were included because, according to one group, "an external speedometer… Was very strong in increasing trust. When a pedestrian has the power to see how fast the [level-5 ADS] is going and can make an informed decision on when to cross based on their own opinion of the situation, it increases [the pedestrian's] trust in the product."

The students also often included lights in their displays, using different colors to signal braking, accelerating, and stopping, which, when combined with the level-5 ADS's

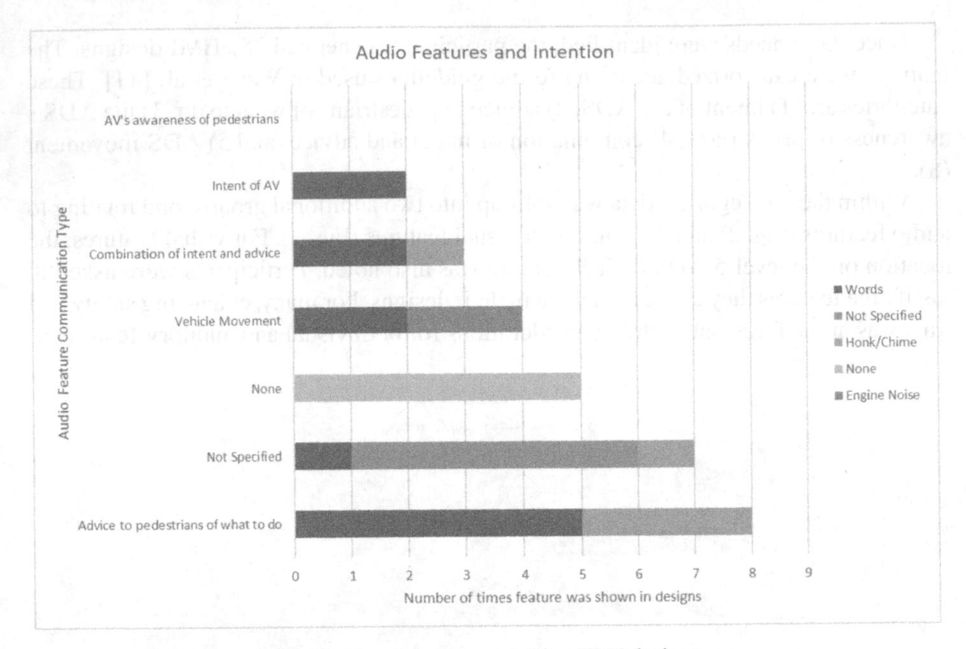

Fig. 3. Aural features used in eHMI designs.

movement, would "enhance safety because the device will communicate to pedestrians when the [level-5 ADS] is in a safe position for the pedestrian to cross." Another group used red LED lights in the shape of an octagon, using a "universally understandable and recognizable" symbol of a stop sign to give the pedestrians information about the level-5 ADS's movements.

Another commonly included visual feature was a countdown timer that showed how many seconds before the ADS stopped. The reason for including a timer was to "enhance safety and trust because the pedestrian [would] be able to see how long they have before the [level-5 ADS] approaches, increasing their confidence that the [level-5 ADS] will stop at that time."

Participant groups also often included level-5 ADS labels somewhere on the ADS, stating that "having the identification will provide more [information] to the pedestrians about the scenario and help increase [their] awareness."

Most visual features were placed on the windshield, grille, hood, or side mirrors of the level-5 ADS. The groups placed various features in these locations, allowing visual features "to be seen from either side of the [level-5 ADS]."

Most groups using audio features mentioned using sounds different from honking "differentiate the…sound from other [traffic]." Another group favored the use of a jingle as the ADS approached, stating that "a jingle would make the sound distinguishable in a busy street, rather than being mistaken for a random… Horn." Other groups had speakers on the level-5 ADS "[announce] when the [level-5 ADS] is starting and stopping" using spoken words. Their justification for the feature was that it would "enhance trust with the pedestrian since the speaker will confirm that the [ADS] has accurately detected them at the intersection." However, the groups did point out that, while language-based eHMI features were the clearest, they were dependent on the pedestrian's language, hearing, and/or sight ability.

Examples of eHMIs designed during the lab study are shown in Fig. 4a and 4b.

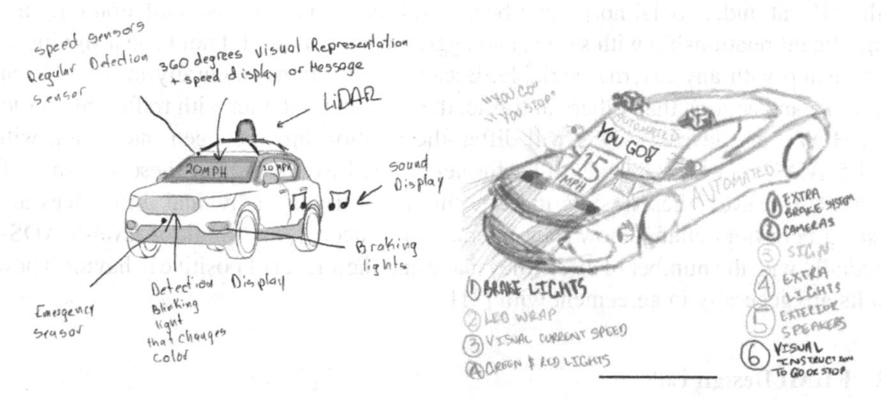

Fig. 4. Examples of eHMI designs created in the lab.

Overall, participants designed eHMIs that were meant to communicate the intention of the level-5 ADS to the pedestrian. The key needs of pedestrians when making safe crossing decisions were reported to be knowledge of the level-5 ADS's speed, if its speed is changing, and if the level-5 ADS has detected the pedestrian. Most often, groups used external speedometers, LED lights, countdown timers, level-5 ADS labels, beeps, jingles, and spoken word.

6 Discussion

The goals of this study were to understand pedestrians' acceptance of and intention to cross the road in front of level-5 ADSs, as well as identify what eHMI features pedestrians most preferred. In this section, the results from the survey analysis and eHMI design lab will be examined and interpreted.

6.1 Theoretical Framework

Personal innovativeness was shown to have a significant relationship with trust, perceived behavioral control, attitude, social norm, and compatibility. Similar to the results in this study, Deb et al. [25] and Kaye et al. [15], it was found that increased levels of personal innovativeness corresponded to increases in pedestrians' acceptance of level-5 ADSs. Deb et al. [9] found that while personal innovativeness did not have an effect on what eHMI features pedestrians preferred, it did have an effect on the amount of time they spent waiting before crossing in front of level-5 ADSs; those with higher personal innovativeness scores showed a lower waiting time before beginning to cross.

Some General Pedestrian Behaviors (GPBs), though, did influence the acceptance factors and intention to cross. Of the five GPB factors, error and positive behavior had the highest number of statistically significant relationships with the factors in the model. The error factor had significant relationships with safety, trust, PBC, understanding, and intention to cross; positive behavior had significant relationships with safety, trust, PBC, social norm, and compatibility. Additionally, lapse had significant relationships with PBC, attitude, social norm, and behavioral intention to cross. Violation only had a significant relationship with safety, and aggressive behavior did not have a significant relationship with any external variable. Because pedestrians' naturally make different decisions and actions than others in traffic, the way they interact with traffic infrastructure, HDVs, and level-5 ADSs will differ–the relationships between interaction with level-5 ADSs and GPBs needs to be further studied to understand these specific differences. However, these results indicate that pedestrians' individual differences and existing behaviors change how they interact with, accept, and think of level-5 ADSs, especially with the number of errors they make and their level of positive behavior. These results are generally in agreement with [21].

6.2 EHMI Design Lab

In the eHMI design lab study, most participants combined visual and auditory features in their designs, but overall used more visual features than auditory features in total. They mentioned safety and trust as the most important factors kept in mind while creating the eHMIs. There were 42 total visual features, but only 24 auditory, across all the designs. This is similar to results from other studies, which have shown that visual (including textual) eHMIs are often preferred and easier to understand [8, 9, 16]. The most commonly occurring visual features were external speedometers, LED lights, countdown timers, and level-5 ADS labels. Most visual features were placed on the windshield, though most designs included features in multiple places on the level-5 ADS and were mainly used to communicate information about movement. Eisma et al. [19] also recommends having eHMIs visible from multiple directions. Auditory features, including beeping, jingles, and spoken words, were commonly used to announce to the pedestrian what to do, such as to cross now or keep waiting.

The inclusion of external speedometers and countdown timers suggests that pedestrians want eHMIs to include information about the motion of the ADS. In the literature, it has been shown that pedestrians will first use implicit communication from human driven vehicles and ADSs to understand their intentions, including speed and distance, to begin deciphering the intention of the vehicle [6–8, 14–17, 19]. For example, a pedestrian at an unsignalized crosswalk would first watch an approaching vehicle to see if it is slowing down to a stop to yield, before using explicit communication with the driver, such as eye contact or a hand gesture, to confirm that it is safe to cross. The literature has shown that this is true for pedestrian interaction with eHMIs on level-5 ADSs as well, where they first collect implicit information, then use the eHMI to verify the vehicle's intention [7, 11, 12, 17, 18, 22]. Including features in an eHMI that quantify implicit information from vehicles, such as an external speedometer showing the speed of the car decreasing, may help pedestrians have more concrete knowledge about the ADS's motion and intention and more quickly verify if it is yielding to them or not. Participants stated that they included these visual features because knowing the speed of the ADS and how long before it arrived at their position would allow them to make safer crossing decisions.

Many eHMI designs also included a label, decal, or other element that declared that the driving system was a level-5 ADS. This could suggest that they would feel more comfortable crossing in front of the ADS if they specifically knew that the driving system was automated. In the literature, pedestrians have been shown to want to know when driving systems are automated, as it gives them a better understanding of how to interact with the ADS [6, 12]. Participants reported that they included the level-5 ADS sign to increase the pedestrian's awareness, however, in the literature, pedestrians have been shown to assume that level-5 ADSs drive more conservatively and will always yield to pedestrians, so including an explicit label could have the potential of encouraging pedestrians to take more risky actions when deciding whether or not to cross in front of a level-5 ADS, as they may assume it will stop for them no matter who has the ROW.

Participants also included audio features, though largely as additions to information given though the visual features or through the ADS's motion, such as playing a jingle or beeping when the level-5 ADS was slowing down, rather than being the main method of giving pedestrians information about its intentions.

The eHMIs that participants designed used combinations of both visual and auditory features to give the pedestrians plenty of information to make a safe crossing decision. However, many groups also included manual override features for emergency situations, suggesting that they may not fully trust level-5 ADS to keep pedestrians safe, and may not feel fully comfortable around them. The students participating in the lab were introduced to level-5 ADSs, which were specified as designed to operate without any input from a human driver, but some groups still included these measures as an additional fail-safe.

Despite significant contributions, this study faced several limitations. First, the sample size for both the survey and eHMI design activity might not be enough for generalization. In the future, studies ought to use large and diverse sample groups that better represent all road users, especially when interacting with eHMIs, whose visual and aural features can be dependent on sight, hearing, language since they may impact pedestrians' preferences of eHMI features [6, 8, 14, 16, 20].

Additionally, participants did not have enough real-life experience interacting with level-5 ADSs. While level-5 ADSs are currently limited in their use in real life, future studies could utilize virtual reality, video, or text-based interaction scenarios, giving survey respondents a better idea of what interacting with a level-5 ADS will be like [7, 9, 13, 16, 18, 23].

Despite these limitations, this study provides valuable information about pedestrians' acceptance of level-5 ADSs and their preferences for eHMIs.

7 Conclusion

Level-5 ADSs [2] have the potential to improve the current traffic system, making conditions safer for vulnerable road users like pedestrians by eliminating driver human error. However, they can only be well integrated into the system if pedestrians can accept them and efficiently interact with them, especially in risky scenarios such as crossing uncontrolled intersection in front of a level-5 ADS. Level-5 ADSs have no ability to communicate directly with pedestrians, unlike how a human driver may use eye contact or hand gestures, but pedestrians must still understand their intentions to facilitate safe road crossing decisions.

In this study, the factors that impact pedestrians' acceptance of level-5 ADSs, as well as their behavioral intention to cross the road in front of them, were explored through an expanded model based on the TPB. In addition to the TPB base factors of perceived behavioral control, attitude, and social norm, we included background factors of gender and previous knowledge of level-5 ADSs, general pedestrian behaviors, personal innovativeness, and the acceptance factors. A survey was created based on this model, and analysis of the results revealed that while the background of participants did not impact acceptance of intention to cross, general pedestrian behaviors and personal innovativeness did. Error, lapse, and positive behavior had the most significant relationships with the external factors of safety, trust, compatibility, and understanding. Personal innovativeness had a significant effect on trust, perceived behavioral control, attitude, social norm, and compatibility.

This study also explored pedestrians' preferences for eHMIs designs, conducting a lab activity. In the study, 25 eHMI designs were generated, with most using a combination of visual and aural features to facilitate communication between the pedestrian and the level-5 ADS. Most visual features were communicating information about ADSs movement, such as speed, while auditory components gave advice to pedestrians of what to do, such as if they ought to cross or yield. Most visual components were located on the windshield of the ADS. These features were included in eHMI designs with the intention of helping pedestrians understand the intention of level-5 ADSs, and cross in front of them efficiently. When participants were designing the eHMIs, they were mainly concerned with increasing levels of perceived safety and trust, and wanted to keep the eHMI designs straightforward while still being informative.

Overall, this study describes the key factors important to pedestrians' acceptance of level-5 ADSs and intention to cross the road in front of them. It also found several key visual and auditory features pedestrians wanted to have included in eHMIs, and why. This study can be used as a foundation to explore other scenarios that researchers have not yet explored regarding level-5 ADSs and pedestrian behavior.

Disclosure of Interests. The authors have no competing interests to declare that are relevant to the content of this article.

References

1. Traffic Safety Facts 2021 Data. https://crashstats.nhtsa.dot.gov/Api/Public/ViewPublication/813458
2. Shi, E., Gasser, T.M., Seeck, A., Auerswald, R.: The principles of operation framework: a comprehensive classification concept for automated driving functions. SAE Int. J. Connect. Automat. Veh. **3**, 27–37 (2020). https://doi.org/10.4271/12-03-01-0003
3. NHTSA Estimates for 2022 Show Roadway Fatalities Remain Flat After Two Years of Dramatic Increases|NHTSA (2022)
4. Singh, H., Kushwaha, V., Agarwal, A.D., Sandhu, S.S.: Fatal road traffic accidents: causes and factors responsible. J. Indian Acad. Forensic Med. **38**(1), 52–54 (2016). https://doi.org/10.5958/0974-0848.2016.00014.2

5. Haghi, A., Ketabi, D., Ghanbari, M., Rajabi, H.: Assessment of human errors in driving accidents; analysis of the causes based on aberrant behaviors. Life Sci. J. **11**(9), 414–420 (2014)
6. Rasouli, A., Tsotsos, J.K.: Autonomous vehicles that interact with pedestrians: a survey of theory and practice. IEEE Trans. Intell. Transport. Syst. **21**, 900–918 (2020). https://doi.org/10.1109/TITS.2019.2901817
7. Pillai, A.: Virtual reality based study to analyses pedestrian attitude towards autonomous vehicles (2017)
8. Bai, S., Legge, D.D., Young, A., Bao, S., Zhou, F.: Investigating external interaction modality and design between automated vehicles and pedestrians at crossings. In: 2021 IEEE International Intelligent Transportation Systems Conference (ITSC), pp. 1691–1696. IEEE, Indianapolis (2021)
9. Deb, S., Strawderman, L.J., Carruth, D.W.: Investigating pedestrian suggestions for external features on fully autonomous vehicles: a virtual reality experiment. Transport. Res. F: Traffic Psychol. Behav. **59**, 135–149 (2018). https://doi.org/10.1016/j.trf.2018.08.016
10. Rothenbucher, D., Li, J., Sirkin, D., Mok, B., Ju, W.: Ghost driver: a field study investigating the interaction between pedestrians and driverless vehicles, pp. 795–802 (2016). https://doi.org/10.1109/ROMAN.2016.7745210
11. Wang, P., Motamedi, S., Qi, S., Zhou, X., Zhang, T., Chan, C.-Y.: Pedestrian interaction with automated vehicles at uncontrolled intersections. Transport. Res. F: Traffic Psychol. Behav. **77**, 10–25 (2021). https://doi.org/10.1016/j.trf.2020.12.005
12. Habibovic, A., et al.: Communicating intent of automated vehicles to pedestrians. Front. Psychol. **9**, 1336 (2018). https://doi.org/10.3389/fpsyg.2018.01336
13. de Clercq, K., Dietrich, A., Núñez Velasco, J.P., de Winter, J., Happee, R.: External human-machine interfaces on automated vehicles: effects on pedestrian crossing decisions. Hum. Factors **61**, 1353–1370 (2019). https://doi.org/10.1177/0018720819836343
14. Lanzer, M., et al.: Designing communication strategies of autonomous vehicles with pedestrians: an intercultural study. In: 12th International Conference on Automotive User Interfaces and Interactive Vehicular Applications, pp. 122–131. ACM, Virtual Event (2020)
15. Kaye, S.-A., Li, X., Oviedo-Trespalacios, O., Pooyan Afghari, A.: Getting in the path of the robot: pedestrians acceptance of crossing roads near fully automated vehicles. Travel Behav. Soc. **26**, 1–8 (2022). https://doi.org/10.1016/j.tbs.2021.07.012
16. Ferenchak, N.N., Shafique, S.: Pedestrians' perceptions of autonomous vehicle external human-machine interfaces. ASCE-ASME J. Risk Uncert. Eng. Syst. Part B Mech. Eng. **8**, 034501 (2022). https://doi.org/10.1115/1.4051778
17. Dey, D., Matviienko, A., Berger, M., Pfleging, B., Martens, M., Terken, J.: Communicating the intention of an automated vehicle to pedestrians: the contributions of eHMI and vehicle behavior. IT – Inf. Technol. **63**, 123–141 (2021). https://doi.org/10.1515/itit-2020-0025
18. Métayer, N., Coeugnet, S.: Improving the experience in the pedestrian's interaction with an autonomous vehicle: an ergonomic comparison of external HMI. Appl. Ergon. **96**, 103478 (2021). https://doi.org/10.1016/j.apergo.2021.103478
19. Eisma, Y.B., van Bergen, S., ter Brake, S.M., Hensen, M.T.T., Tempelaar, W.J., de Winter, J.C.F.: External human-machine interfaces: the effect of display location on crossing intentions and eye movements. Information **11**, 13 (2019). https://doi.org/10.3390/info11010013
20. Fratini, E., Welsh, R., Thomas, P.: Ranking crossing scenario complexity for eHMIs testing: a virtual reality study. MTI **7**, 16 (2023). https://doi.org/10.3390/mti7020016
21. Deb, S., Strawderman, L., DuBien, J., Smith, B., Carruth, D.W., Garrison, T.M.: Evaluating pedestrian behavior at crosswalks: Validation of a pedestrian behavior questionnaire for the U.S. population. Accid. Anal. Prevent. **106**, 191–201 (2017). https://doi.org/10.1016/j.aap.2017.05.020

22. Zhao, X., Li, X., Rakotonirainy, A., Bougeois-Bougrine, S., Delhomme, P.: Predicting pedestrians' intention to cross the road in front of automated vehicles in risky situations. Transport. Res. Part F: Traffic Psychol. Behav. **90**, 524–536 (2022). https://doi.org/10.1016/j.trf.2022.05.022
23. Jayaraman, S.K., Tilbury, D.M., Jessie Yang, X., Pradhan, A.K., Robert, L.P.: Analysis and prediction of pedestrian crosswalk behavior during automated vehicle interactions. In: 2020 IEEE International Conference on Robotics and Automation (ICRA), pp. 6426–6432. IEEE, Paris (2020)
24. Das, S.: Autonomous vehicle safety: Understanding perceptions of pedestrians and bicyclists. Transport. Res. F: Traffic Psychol. Behav. **81**, 41–54 (2021). https://doi.org/10.1016/j.trf.2021.04.018
25. Deb, S., Strawderman, L., Carruth, D.W., DuBien, J., Smith, B., Garrison, T.M.: Development and validation of a questionnaire to assess pedestrian receptivity toward fully autonomous vehicles. Transport. Res. Part C: Emerg. Technol. **84**, 178–195 (2017). https://doi.org/10.1016/j.trc.2017.08.029
26. Ajzen, I.: The theory of planned behavior. Organ. Behav. Hum. Decis. Process. **50**, 179–211 (1991). https://doi.org/10.1016/0749-5978(91)90020-T
27. Davis, F.D.: Perceived usefulness, perceived ease of use, and user acceptance of information technology. MIS Q. **13**, 319 (1989). https://doi.org/10.2307/249008
28. Venkatesh, M., Davis, D.: User acceptance of information technology: toward a unified view. MIS Q. **27**, 425 (2003). https://doi.org/10.2307/30036540

Understanding Commuter Information Needs and Desires in Public Transport: A Comparative Analysis of Stated and Revealed Preferences

Anouk van Kasteren(✉), Marloes Vredenborg, and Judith Masthoff

Utrecht University, Utrecht, The Netherlands
{a.vankasteren,m.t.r.vredenborg}@uu.nl

Abstract. This paper explores commuters' stated- and revealed preference information needs and desires in public transport through a multi-method study. A survey with 286 participants uncovers stated preferences, while a diary study with 31 participants provides insight into the revealed preferences. Key findings include a comprehensive overview of travellers' information needs and desires, offering insights to improve current traveller information systems. Moreover, the results show differences between real-time information needs and those recalled from memory. During disruptions, important information includes information about the duration, cause of disruption, consequences for the journey and alternatives, while information about potential disruptions, schedule, journey planning and interchanges are important for journeys without disruptions. Another important finding was the significant difference between the needs and desires during a regular and disrupted journey.

Keywords: Public Transport · Information Needs · Commuters · Disruptions · Diary Study

1 Introduction

In an attempt to create a more sustainable society, governments strive to reduce carbon emissions. One strategy is to promote the use of public transportation (e.g. [16]), a sustainable alternative in contrast to private car usage. Unfortunately, since the Covid-19 pandemic, there has been a decrease [23] and behaviour change [20] in the use of public transport. A substantial proportion of these travellers have resorted back to private vehicle usage [17,28]. In 2021, commutes make up a large proportion of the distance travelled in the Netherlands. 70.6% of this is done by private car [8]. Therefore, encouraging commuters to make more use of public transport could have a significant impact.

Yet, for this to be effective, public transport must be accessible and offer a convenient and pleasant travel experience [37]. Planning a public transportation

A. van Kasteren and M. Vredenborg—The two authors contributed equally to this paper.

H. Krömker (Ed.): HCII 2024, LNCS 14733, pp. 83–103, 2024.
https://doi.org/10.1007/978-3-031-60480-5_5

journey can pose challenges, especially when navigating unexpected disruptions [1]. Clear information provision is crucial for travellers to successfully navigate public transportation [21]. Furthermore, providing the correct information to the right person at the right time can significantly improve the travel experience [7,33], particularly during disruptions.

Understanding the "Who-When-Where-What" of information needs and desires of passengers is vital for achieving better information provision [24]. There is limited work on the real-time information needs and desires of commuters. Therefore, this research enriches the current knowledge by conducting an exploratory study employing a stated- (survey) and a revealed (diary-study) preference method. Moreover, it explores the potential for tailoring travel information by assessing relationships between different information types and demographic or trip-related factors.

The remainder of this paper is structured as follows. Section 2 outlines the related work. Sections 3 and 4 describe the method and results. Section 5 discusses the results, practical implications, and possible future work. Lastly, Sect. 6 concludes the paper.

2 Related Work

Numerous studies have emphasised the value of providing relevant travel information in public transport. Often, a differentiation is made amongst different stages of the travel (e.g. pre-trip, during travel, last mile). For instance, Hörold et al., [19] created a framework for identifying passenger information needs based on the tasks and steps within the travel chain. Here, as part of their research, they define that information about disturbances is necessary at all stages of the travel.

Furthermore, Berggren et al. [5] emphasize the importance of having access to pre-trip information, particularly for long journeys, to pre-plan and thereby minimise waiting times. Others have shown the importance of relevant information at different journey segments for passengers' perceived service quality and satisfaction of light rail in Madrid [33]. Farag and Leyons [15] showed that pre-trip information use is related to travel behaviour and socio-demographics. They suggest that the use of information is more related to "the person" than to the specific trip, indicating the potential of personalized, relevant information.

Additionally, how and where to best communicate the information has also been studied. For instance, (schematic) maps are found to be more efficient in terms of comprehending travel information compared to lists [4]. Work by Mulley et al. [27] shows that travellers use different information sources across various stages of the journey. They also found differences in used information sources between bus and train travellers. Moreover, real-time information through mobile devices has been associated with reductions in overall travel time, more transit use, and increased satisfaction [6].

Studies have been conducted to determine the specific type of information that travellers require. Tang et al. [35] investigated the information needs of travellers in Chengdu, China, focusing on what, when, and how much information is necessary. The results showed that most participants value information that can help people save travel time and make the trip easier. They also showed that people in different socio-economic segments require varying quantities of information. Papangelis et al. [29] examined the information needs of bus passengers in rural areas during disruptions. Passengers generally seek information about the severity and impact of disruptions pre-trip and pre-disruption. After they have been warned, passengers prefer to receive real-time updates on the status of their transportation and alternative transport options. In a diary study conducted by Chen and Qi [10], the information needs of leisure travellers were explored. The study identified eight types of questions, three types of intents, and eighteen topics. The study's results revelead travellers most want the "where," "what," and "how" type of information.

Lastly, passengers' information-seeking behaviour has been linked to factors related to trip frequency, socio-demographics, trip context and information sources [40]. This study, however, merely looked at whether information was sought, they did not look at the effects of specific information content.

Previous studies have left gaps in understanding the differences between information needs and desires of commuters in disrupted and regular journeys, and the relationship between specific information content and demographic or trip-related factors. Our research aims to address these gaps by identifying a comprehensive set of information needs and desires and comparing their occurrence between disrupted and regular journeys. We also conduct an association analysis and compare the results of stated- versus revealed preference studies.

3 Method

We conducted a survey and a diary study to explore commuters' information needs and desires in public transport. The survey allowed us to understand better commuters' stated information needs and desires. The diary study provided more insight into the revealed information needs and desires. Both studies were approved by the Utrecht University Research Institute of Information and Computing Sciences based on an Ethics and Privacy Quick Scan.

3.1 Survey

We created an exploratory survey to capture the **stated** information needs and desires of commuters in a door-to-door public transport journey. We asked participants to think about their past travels. We chose a survey as a research method, as it allowed us to capture a broad range of information needs and desires. In the following subsections, we will describe the method in detail.

Sample. 286 participants, see Table 1 for all demographics, were recruited using the traveller panel of the Dutch Railways[1]. This panel represents a heterogeneous

[1] https://www.nspanel.nl/.

group of public transport travellers in the Netherlands and encompasses over 80.000 members. Participants were required to use public transport for work or school commuting. We aimed to have a diverse sample in terms of gender, age and province of residence. Of the 286 participants, 88% use the train, 25% the bus, 13% light rail (metro and tram), and 7% shared bikes. Informed consent was obtained before the survey was conducted.

Protocol. Participants received an online survey to express their stated information needs and desires. We started the survey by asking the participants to think about their public transport journeys for work or study from door to door. We told them that we would ask questions about their information needs and desires, and the information quality in these journeys in both regular and disrupted situations. Upon completion, participants were thanked for participating and asked about their willingness to participate in a two-week diary study. Completing the survey took approximately eight minutes.

Survey Questions and Measurements. The survey questions asked participants about (1) their information needs and desires, (2) the information quality, and (3) the journey quality during their door-to-door public transport journey in regular and disrupted situations.

Information Needs and Desires. To assess participants' *information needs* during their door-to-door public transport journey, we asked open-ended questions about the information they need to have on their journey with public transport to complete their journey successfully. Additionally, we asked questions about what information they would like to have to enhance their travel experience, also referred to as *information desires.*

Information Quality. Next, participants rated the information on availability, findability, clarity, correctness, completeness and communication satisfaction using a 5-point Likert scale for both a regular and disrupted journey.

Journey Quality. Finally, participants rated their average work/study-related trip by public transport, considering the entire door-to-door journey, on a scale from 1 (very poor) to 10 (excellent).

Analysis. The analysis of the results is exploratory, aiming to identify travellers' information needs and explore possible associations to guide further research. The analysis is threefold.

First, the survey data was analyzed in Nvivo [32]. Through inductive coding themes were identified. Through a bottom-up approach, the information needs and desires were categorised resulting in, from now on referred to as, information types. Crosstabs were generated to check the presence of information types in regular and disrupted journeys. An information type was marked as 1 if mentioned and 0 if not. The crosstab was exported and merged with the original dataset. Summary statistics were generated and a frequency analysis was performed. A Cochran's Q test[2] was performed to determine the relative relevance of the top 5 information types for both regular and disrupted journeys.

Second, a McNemar [26] test[3] was performed to analyze the difference in proportions between regular and disrupted journeys per information type. The p-values were adjusted using the Bonferroni correction due to multiple comparisons [14]. Demographics and trip-related data were tested for associations with each information type using χ^2 tests [30] and Cramer's V effect size [13]. If an association was found, a pairwise post-hoc test was performed with Bonferroni correction for multiple comparisons.

Lastly, the information quality was analysed by combining the variables availability, clarity, correctness, completeness, and findability into a new variable called quality (Cronbach $\alpha = 0.93$). Averages were computed and compared between regular and disrupted journeys using a Wilcoxon signed rank test [39] and a Rosenthal correlation effect size [34].

3.2 Diary Study

We created a two-week diary study to understand the **revealed** information needs and desires of commuters in a door-to-door public transport journey. This allowed us to collect real-time, longitudinal insights. Before the actual diary study, two pilot studies were conducted. After each pilot, the diary study protocol was assessed and revised based on the participants' feedback. In the following sections, we describe the method in more detail.

Sample. Participants were recruited via two channels. First, participants were asked at the end of the survey (see Sect. 3.1) if they would also be willing to participate in a diary study. Participants who showed interest in participating were contacted via e-mail. Second, to ensure a more diverse sample, we recruited extra participants through PanelClix[4]. An additional eligibility requirement was that participants make use of at least one other transportation mode besides the train. These participants were compensated with a ten-euro gift card. In total, 31 participants completed the diary study; 21 were recruited via the first channel and 10 via the second. Table 1 shows the demographics of the sample. Informed consent was obtained before the diary study was conducted.

Protocol. Before conducting the diary study, all materials were prepared: an e-mail invitation, an informed consent form, an instruction manual, (video)call protocol and a diary survey programmed in Qualtrics.

After participants expressed their willingness to participate in the diary study, we contacted them via e-mail and provided them with all details of the

[2] Cochran's Q was selected due to the paired nature of the data.

[3] This test was performed as the assumption of independence was violated.

[4] www.panelclix.nl.

Table 1. Demographics of the survey and diary sample.

		Survey		Diary	
		N	%	N	%
Gender	Male	163	57	17	55
	Female	116	41	13	42
	Other	4	1	1	3
	(Unknown)	3	1		
Age	18–35	28	10	5	16
	36–55	68	24	7	23
	55+	175	61	19	61
	(Unknown)	15	5		
Province of residence	Noord-Holland	10	3	4	13
	Zuid-Holland	18	6	4	13
	Friesland	35	12	2	6
	Noord-Brabant	33	12	2	6
	Gelderland	66	23	7	23
	Overijssel	40	14	5	16
	Limburg	77	27	3	10
	Flevoland			2	6
	Utrecht			1	3
	Groningen			1	3
	(Unknown)	7	2		
Travel frequency	4 days per w eek or more	82	29	14	45
	1–3 days per week	89	31	13	42
	1–3 days per month	53	19	2	6
	Less then 1 day per month	56	20	1	3
	(Unknown)	6	2	1	3

study. We provided them with the instruction manual and an invitation to a (video) call scheduled the week before they started the study. During the video call, we introduced ourselves and asked the participants to introduce themselves, including their age, area of residence, travel frequency, and standard commute. Their standard commute was explained to them as the route that they usually take as a home-work/home-study trip, including departure and arrival points, transfer locations and the means of public transport. Subsequently, we addressed any questions participants had about the manual, and we went through the diary study to ensure clarity and discuss potential issues.

On the Friday preceding the study's start, we sent them a participant ID and a link. They were instructed to complete the diary survey after each commute for two consecutive weeks, recording information separately for each journey. Completing the diary survey took approximately 3 min per entry. Throughout these two weeks, we provided reminders and support via WhatsApp, SMS, or email according to their preference. After two weeks, participants were asked if they had any additional remarks, and we thanked them for their participation.

Diary Survey Questions and Measurements. The diary survey asked participants about public transport mode choices and disruptions experienced before or during the journey. Subsequently, participants were directed to specific questions based on whether their journey was regular or disrupted.

Regular Journey. When asking participants about their regular journeys, we inquired about additional information they desired beyond basic requirements. Questions included whether the information was desired before departure or during the journey and in relation to what means of transport. An open-ended question asked participants to specify the type(s) of information sought.

Disrupted Journey. In the case of a disrupted journey, we asked participants to specify the mode of transportation affected. Thereafter, open-ended questions were asked, outlining the timing, nature, impact, and subsequent actions in response to the disruption (e.g., wait, detour, not travel). Participants were further asked to specify the information needed during the disruption, with eight predefined categories[5]: (1) cause of disruption (2) disruption duration (3) consequences for the schedule (4) consequences for the journey (5) alternatives (6) tickets (7) location and navigation (8) facilities. Moreover, there was an open-ended option where participants could describe additional information needs. In case of multiple disruptions, participants answered these questions for each transportation mode separately.

Following either the regular or disrupted journey questions, participants were asked to rate the availability of information on a three-point scale (not available - partially available - fully available). If information was partially or fully available, we asked them to rate it on completeness, findability, clarity, and correctness using a 5-point Likert scale. Participants were also asked to rate their satisfaction with information communication on a 5-point Likert scale, accompanied by an open-ended question asking where they found the information. These questions were asked separately for each information category in case of disruptions. For regular journeys, these questions were asked once for the answers in the open-text field.

At the end of the diary survey, all participants were asked to rate their door-to-door journey, on a scale from 1 (very poor) to 10 (excellent).

Analysis. Two datasets were created from the raw data exported from Qualtrics. One provided an overview of transport modes, if there were disruptions or information needs and desires, and demographics. The second dataset focused on information needs, with each line corresponding to a single need and quality score. The open-text entries were categorized with a bottom-up approach. The final dataset contained the elaborated information needs' answers as well as an information category variable. Additionally, the dataset contained the disruption and demographics data. The resulting information types were merged with those identified from the survey results to create a comprehensive list.

[5] Categories were based on previous results.

A descriptive analysis was performed on the overview dataset to gain insights into the frequencies of disruptions and information needs or desires. A Cochran-Mantel-Haenszel (CMH)[6] test [11] was performed to test the association between disruptions and the need for information.

Next, a descriptive analysis was performed on the information need dataset to identify the top 5 information types for regular and disrupted journeys. Associations between the top 5 information needs and trip-related data were tested by CMH tests. Additionally, the information quality was analysed. A new quality variable was created by combining completeness, findability, clarity, and correctness (Cronbach's $\alpha = 0.91$). Moreover, for each information type it was analyzed how often it was not, partially, or fully available.

4 Results

The results will be addressed in three parts. First, the survey results will be discussed, and next the diary study results. Third, the results of the two studies are compared.

4.1 Survey

Participants often provided short answers to the open-ended questions such as "*arrival time*" or "*duration of the disruption*". Some provided longer, more elaborate answers. For example, one participant responded to what information they need for a regular journey:

"*What time does my train leave, which transfer do I have to make, which platform do I arrive on and where does my next train go from? How long is my transfer time?*";

and for a disrupted journey: "*When does the next train leave, which platform does this train leave from? What time does this train arrive and what does this mean for my transfer?*".

The four themes identified from these answers were information types, when and where to get information, communication channel, and information quality.

When, Where, and How to Get Information. Not all participants mentioned these themes in their answers. 28% of the participants mentioned the communication channel to get information. They mentioned travel planner apps (18%), announcements (9%), signs and screens (7%), and a website (3%). 17% of participants mentioned when or where to get information. Among the results were at the station (7%), on the platform (2%), at the stop (1%), and in the vehicle (7%). The information mentioned at the station or on the platform often regarded the travel times or possible changes to these. Arrival times and information about interchanges were among the answers mentioning "in the vehicle".

[6] This test was selected due to the repeated observations in the data.

Information Types. The coded information needs were classified and resulted in 17 overarching information types. Table 4 provides an overview of all identified types and their descriptions.

A frequency analysis of the information types clearly shows the three most frequently mentioned information types. These are journey planning (n = 193), (possible) disruptions (n = 192), and alternatives in case of disruptions (n = 180). One recurring theme within journey planning was information about time (e.g. departure and arrival time, waiting time, or travel duration). In total, 161 participants mentioned an information type related to travel time. Other examples of themes mentioned related to journey planning were the departure station or platform and the price of the journey. The information type disruption regarded information about possible delays, cancellations or changes. Alternatives considered either information about alternative routes or replacement transport offered by the transportation carrier.

The top 5[7] information types disruptions duration are: (1) Alternatives, (2) disruption duration, potential disruption, and journey planning, and (3) cause of the disruption. In contrast, the top 5 (see footnote 7) for a regular journey are: (1) potential disruptions, journey planning, (2) interchange, alternatives, and crowdedness.

Figure 1 (left-hand side) provides an overview of the frequencies of the top 10 information types in case of a regular or disrupted journey. This overview shows the discrepancies in the frequencies between regular and disrupted journeys. For example, as can be expected, information on alternatives or disruption duration is much more frequently mentioned in a disrupted compared to a regular journey. The opposite is true for journey planning and (possible) disruptions. More specifically, a McNemar [26] test revealed significantly differing distributions in participants mentioning information types between disrupted and regular journeys, except for 'location & navigation' and 'facilities' categories. Table 2 shows these results. Journey planning was mostly mentioned for a regular journey. However, looking at the sub-types in journey planning, the following were mentioned more often in a disrupted journey: the next option (100%) and waiting time (75%).

Associations. Due to the exploratory nature of this research, a number of associations between demographics or trip-related factors and the different information types were tested. This section will discuss the noteworthy findings. Table 3 presents a complete overview of the effects identified.

Several age-related effects were found. For example, there is a significant relation with a small effect size between age and information on crowdedness ($\chi^2(2, N = 286) = 10.4$, $p = 0.006$ $V = 0.20$). A post-hoc test showed the 55+ category to have a difference in frequency (9%) compared to 18–34 (27%, p-adjusted = 0.046). Similar trends, older participants needing the information less frequently, were also observed for schedule, location&navigation, composition vehicle, journey planning, and disruption duration. The largest effect size was observed for journey planning (V = 0.24).

[7] Based on a Cochran's Q [11] analysis.

For regular journeys, gender was associated with information on journey planning ($\chi^2(1, N = 279) = 4.3$, $p = 0.037$) with a small effect size $V = 0.13$). Women (74%) mentioned journey planning more frequently than men (63%).

A significant association with a large effect size was found between region and information on disruption duration ($\chi^2(5, N = 286) = 19.6$, $p = 0.001$ $V = 0.27$). A pairwise post-hoc showed that participants with postal codes 6000–6999 mentioned it significantly more often compared to 7000–7999 (p = 0.03) and 8000–8999 (p = 0.04).

The travel frequency was found to have a significant association with the information types alternatives, crowdedness, potential disruptions, and duration of the disruption. The largest effect size was observed for potential disruptions $\chi^2(3, N = 286) = 17.9$, $p =< 0.001$ $V = 0.26$. A pairwise post-hoc test with Bonferroni correction revealed significant differences between 1–3 days per month and 4 or more days per week (p = 0.022), 1–3 days per week and less than 3 times per month (p = 0.031), and 4 or more times per week and less than 3 times per month (p = 0.002). Participants who travel more reported more often to need or want information about possible disruptions. Similar effects were observed for the other information types with significant associations with travel frequency.

Last, the association between departure station size and the information types was tested. In a disrupted journey, a significant effect was observed with the information type interchange ($\chi^2(1, N = 286) = 12.5$, $p =< 0.004$ $V = 0.24$). Participants who generally depart from a small station mention information about an interchange more often than those from a large station. Additionally, a significant association was found between station size and journey planning ($\chi^2(1, N = 286) = 6.6$, $p =< 0.010$ $V = 0.17$). Again, the participants departing from small stations mentioned this information type more often.

Table 2. Results of multiple McNemar tests on the difference in information need frequencies between a disrupted and regular journey. P-values are adjusted with a Bonferroni correction.

Category	χ^2	p-value
Alternative	117.5	<0.001
Availability	8.5	0.023[a]
Interchange	31.7	<0.001
Schedule	12.9	0.003
Crowdedness	25.3	<0.001
Location & Navigation	0.3	1
Journey planning	69.3	<0.001
Composition vehicle	16.4	<0.001[a]
(Potential) disruptions	81.1	<0.001
Facilities	6.7	0.074[a]

[a] The exact binomial test is done when in the contingency table, the sum of not mentioned in a regular journey and not mentioned in a disrupted journey < 25

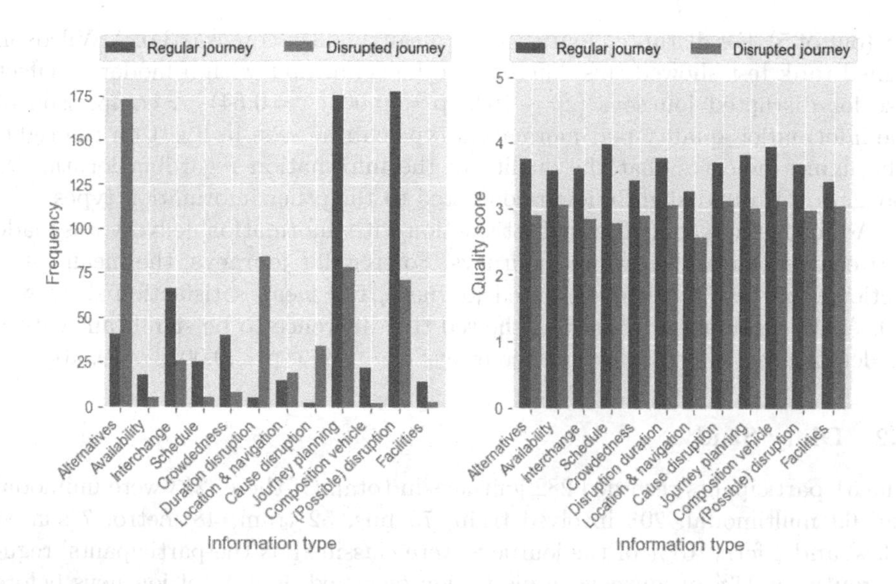

Fig. 1. Left: Comparing frequencies of participants mentioning information types between a regular and disrupted journeys. Right: Comparing the quality score for each information type between regular and disrupted journeys.

Information Quality. The information quality ratings were compared between regular and disrupted journeys. For regular journeys, the mean quality score was

Table 3. Associations between groups and information types based on multiple χ^2 tests [31]. Values correspond to the effect size Cramer's V. P-values < 0.05 are marked with *, p-values < 0.01 are marked **, and p-values < 0.001 are marked ***. are based on Cell shading is based on effect sizes; darker equals a larger effect size. Associations only present for regular journeys are marked α, association only present for disrupted journeys are marked with β.

	Effect size				
	Demographics			Travel-related factors	
	Age	Gender	Region	Travel freq	Station size
Alternatives				0.20*	
Interchange					0.16*
Schedule	0.20**				
Crowdedness	0.20**			0.20*	
Location&Navigation	0.19**				
Journey planning	0.24***	0.13*$^{\alpha}$			0.17*$^{\beta}$
Composition vehicle	0.22**				
Potential disruption				0.26***	
Disruption duration	0.15*		0.27**	0.20*	

3.8 (out of 5). For disrupted journeys, the mean quality score was 3.0. A Wilcoxon signed rank test showed this difference to be significant with a moderate effect size for disrupted journeys (z= −10.9, p < 0.001, r = 0.64). A comparison of the information quality per information type can be seen in Fig. 1 on the right. The figure indicates that the quality of the information regarding location & navigation is rated slightly lower compared to the other information types.

Additionally, a comparison of satisfaction with information delivery was made between regular and disrupted journeys. For regular journeys, the mean satisfaction score was 3.8. For disrupted journeys, the mean satisfaction score was 3.0. A Wilcoxon signed rank test showed the difference to be significant with a moderate effect size for disrupted journeys (z = −9.5, p < 0.001, r = 0.56).

4.2 Diary Study

The 31 participants recorded 282 journeys in total. Of these, 213 were unimodal and 69 multimodal. 208 involved train, 73 bus, 52 tram, 48 metro, 7 shared bikes, and 2 ferry. 67% of the journeys were classified as the participants' regular route. In 24% of journeys, a disruption occurred, in 15% of journeys before departure and 15% during the journey. For 33% of journeys (N = 93) an information need/desire was registered. This resulted in a total of 186 registered information needs/desires. Due to the setup, most participants used the predefined information types when indicating their information needs. Sometimes, participants elaborated on their answer, for example, "*location first class*", "*potential disruption*", "*coffee*", "*location silent compartment*". Comparing regular with disrupted journeys, an information need or desire was more often reported in case of disruption (90.6% vs 15.3%). The results of a CMH test show a significant association (OR = 47.3, p < 0.001) between the two variables. The constant odds ratio test showed no differences between individual participants.

Information Types. Results from the diary study show that for disrupted journeys the top 5 information types are: (1) Disruption duration, (2) Consequences journey, (3) Cause of disruption, (4) Alternatives, and (5) Consequences for the scheduling. For regular journeys, there was predominantly a need for information about potential disruptions. Additional information types mentioned were about the schedule, facilities, and journey planning. Figure 2 (top-left) shows the frequencies of the different information types for regular and disrupted journeys.

Associations. CMH tests [25] were done on the top 5 information types in case of disruption to find potential associations in the data. Several associations were found. Disruption type (delay or cancellation) is significantly associated with the information types alternatives ($\chi^2_{MH} = 74.2, p < 0.001$, OR = 0.42), disruption duration ($\chi^2_{MH} = 101.2, p < 0.001$, OR = 2.30) and cause of disruption ($\chi^2_{MH} = 18.6, p < 0.001$, OR = 1.45). In case of a delay, duration and cause were mentioned more often. In case of cancellation, alternatives were mentioned more often.

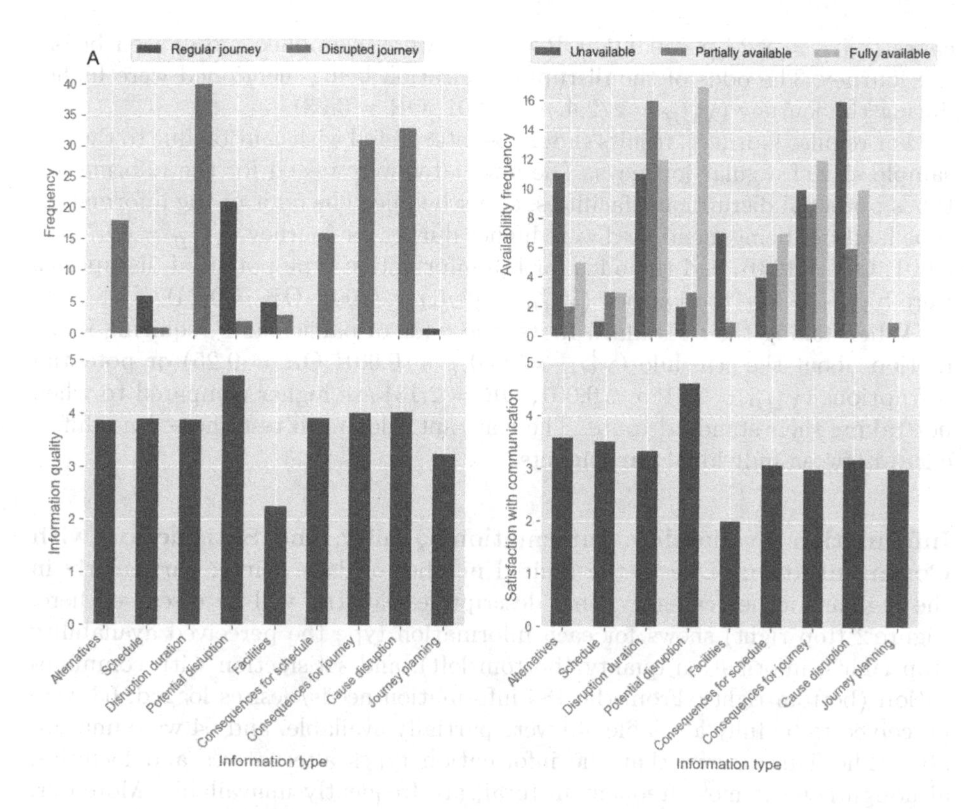

Fig. 2. Top left: frequencies of the information types mentioned in regular and disrupted journeys. Top right: frequencies of availability of the different information types, divided into unavailable, partially available, and not available. Bottom left: the mean information quality score per information type. Bottom right: the mean satisfaction score of how the information is communicated per information type.

Significant associations between the station size and information types showed that participants with large departure stations mentioned consequences for the scheduling ($\chi^2_{MH} = 38.9, p < 0.001$, OR $= 0.45$), consequences for the journey $\chi^2_{MH} = 80.1, p < 0.001$, (OR $= 0.36$), and alternatives (OR $=$ $\chi^2_{MH} = 27.9, p < 0.001$, 0.50) more often. Participants with small departure stations mentioned the cause of the disruption more often ($\chi^2_{MH} = 51.7, p < 0.001$, OR $= 2.00$).

When taking their standard route, participants mentioned alternatives ($\chi^2_{MH} = 249.0, p < 0.001$, OR $= 0.16$) and consequences for the scheduling significantly ($\chi^2_{MH} = 191.8, p < 0.001$, OR $= 0.20$) less. On the contrary, the disruption duration ($\chi^2_{MH} = 12.8, p < 0.001$, OR $= 1.29$) and disruption cause ($\chi^2_{MH} = 170.5, p < 0.001$, OR $= 2.83$) were mentioned more often when participants took their standard route.

Consequences for the journey ($\chi^2_{MH} = 10.3, p < 0.001$, OR $= 0.78$), consequences for the schedule ($\chi^2_{MH} = 67.8, p < 0.001$, OR $= 0.39$), and disruption cause ($\chi^2_{MH} = 8.27, p < 0.001$, OR $= 0.80$) were mentioned more often before the journey. The odds of the disruption duration being mentioned were higher during the journey ($\chi^2_{MH} = 272.4, p < 0.001$, OR $= 3.33$).

For regular journeys, results should be interpreted with caution due to the low sample size of regular journeys. The associates were tested for the information types potential disruptions, facilities, and schedule. The odds of the information type facilities being mentioned were higher during the journey ($\chi^2_{MH} = 69.7, p < 0.001$, OR $= 12.0$), and the odds for the information type potential disruptions were higher before the journey ($\chi^2_{MH} = 46.4, p < 0.001$, OR $= 0.24$).

When taking their standard route, the odds of participants requiring information about the schedule ($\chi^2_{MH} = 28.0, p < 0.001$, OR $= 0.25$) or potential disruptions ($\chi^2_{MH} = 11.1, p < 0.001$, OR $= 2.13$) are higher compared to when not taking their standard route. The constant odds ratio test showed no differences between individual participants.

Information Availability, Information Quality, and Satisfaction with Communication. Due to the limited number of data points, particularly in the regular journey category, only descriptive statistics will be discussed here. Figure 2 (top right) shows for each information type the perceived availability (top right), information quality (bottom left), and satisfaction with communication (bottom right). From the 186 information needs/desires logged, 68 were perceived to be fully available, 49 were partially available, and 64 were unavailable. The figure shows that the information types alternatives and facilities, although not the most frequent in total, are frequently unavailable. Moreover, the information type disruption cause is also often unavailable. The information type possible disruptions, most often mentioned in regular journeys, is often available. The mean quality score is 4.09; the mean score regarding the satisfaction about the communication is 3.5. Looking at regular and disrupted journeys separately, the mean quality of the available information is 4.46 and 3.96 respectively. However, the communication satisfaction was better in a regular journey (4.32) compared to a disrupted journey (3.25). The figure shows that especially the information type facilities seems to score lower.

4.3 Comparison Survey and Diary Study

Comparing the results of the survey and the diary study (Fig. 3) shows several discrepancies. For example, alternatives and journey planning were mentioned more often in the survey, while disruption duration, cause of disruption, and consequences for the journey and schedule were more frequently mentioned in the diary study.

For regular journeys, the information types found in the diary entries were mainly regarding information about potential disruptions. Responses to the survey were more diverse.

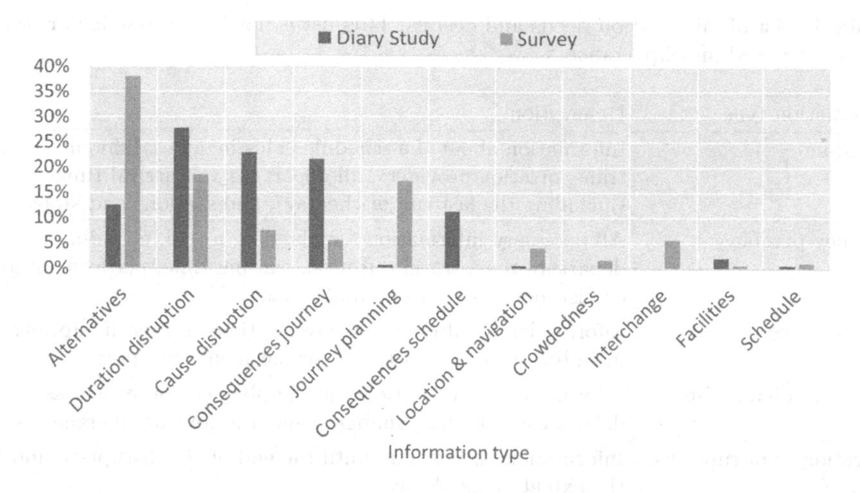

Fig. 3. Comparison of proportional information needs during disruptions between the survey and diary study. Highlighting differences in needs for alternatives, disruption cause, journey planning, and journey consequences.

4.4 Information Types

Based on the coded and categorized answers from the survey, as well as the categorized answers from the diary study, a comprehensive list of travellers' information requirements was created (see Table 4). Its purpose is to provide insights into the various types of information that may be required, as well as what these types of information involve.

5 Discussion

The survey and diary study provided several noteworthy results. During a disruption, travelers' information needs differ significantly from an undisrupted journey. Travelers are more concerned with the consequences of the disruption and how to continue their journey, rather than facilities and crowdedness. Moreover, the type of disruption affects the information needs of travellers. When facing a delay, they want to know the cause and duration. This enables them to make informed decisions, for example, on whether to wait. In the case of cancellations, they are more interested in alternative options.

Interestingly, the survey showed more diverse responses for regular journeys compared to the diary study. An explanation may be that participants in the survey had more time to recall past travels. In the diary study, participants reported on their immediate needs and desires which mostly focused on potential disruptions. Information about, for instance, crowdedness, staff, or composition vehicle were not mentioned. According to the customer wish pyramid defined by van Hagen and van Oort [18], these higher-level information types relate to

Table 4. List of information needs and desires. This list is not in any particular order and is not based on importance.

Information type	Description
Schedule	Information about the schedule. This includes different lines, transportation providers, all departure and arrival times (including the last run of the day), and stations and stops
Journey planning	All necessary information for planning a trip, including departure time, travel duration, waiting time, platform, stop or distance, fastest route and pricing.
Alternatives	Information on alternative travel options during disruptions, including routes, modes, and replacement transport.
Potential disruptions	Information on disruptions on the planned route such as delays, cancellations, changes, constructions or emergencies.
Duration of disruptions	Information on the time until the end of the disruption and the extent of the delay.
Cause of disruptions	The incident that caused the disruption. To create understanding and help evaluate the severity.
Journey consequences	The impact of a disruption on the journey. Including extra travel time, price difference, re-routing, feasibility of the journey and whether one can make it home.
Schedule consequences	The impact of the disruption on the schedule. Includes changes of the departure times, changes of platform, or lines being cancelled.
Crowdedness	The amount of other travellers at the stations and in the vehicles. It also includes the spreading of travellers.
Composition vehicle	Type of vehicle, seat arrangement, compartment arrangement and length of the vehicle.
Location & navigation	Information about the location of vehicles, compartments, free seats, doors, and crowds. (Micro) navigation to and inside places, including traffic.
Interchange	All information regarding interchanges, such as transfer options, number of interchange locations, time available to change, walking distance, and feasibility.
Current time	Information on the current time. For instance, availability on screens, platform clocks or in the vehicle.
Staff	Information about the presence and availability of staff, how to reach them, and ticket control.
Availability	Information about the availability of certain transport modes, including available shared mobility, undisrupted connections, or an available connection late at night.
Seats	Number of (free) seats, seats available in certain compartments or type of seats present.
Parking	The available parking for private transport, such as bikes or cars, and shared mobility.
The weather	The current weather or predicted weather on the travel day. For instance, bike appropriate weather

comfort and experience. Not considering these factors may imply that lower-level needs for reliability, speed, and ease remain unmet.

Several associations were found in the data. Interestingly, older travellers' require less information about the schedule, location & navigation, vehicle composition, disruption duration, journey planning, and crowdedness. This may be due to them relying more on their own knowledge and memory rather than needing information in the moment since they are likely to be more experienced travellers and also grew up in a time with less immediate access to information. More research would be needed on the nature of this association. Furthermore, women want journey planning information more frequently than men, potentially due to higher conscientiousness [38] and neuroticism [12]. However, one must be wary of stereotyping. Travel frequency is associated with the need for information about alternatives, crowdedness, potential disruption, and disruption duration. Interestingly, participants who travel more often want this information more frequently. As for the disruption duration, frequent travellers might be interested in knowing if their following journeys will be possible. Additionally, more frequent travellers are also more aware of potential issues such as disruptions or crowdedness. Moreover, being more familiar with the schedule opens up a mental load to consider additional factors [2,3]. Participants travelling from smaller stations want more information regarding journey planning and interchanges. Smaller stations likely only have regional connections, resulting in journeys being more complex and involving at least one interchange. Lastly, travellers who deviate from their regular route often need assistance with alternatives and consequences in case of a disruption. This could be due to being less familiar with the route and the options in case of a disruption.

Both the survey and the diary study indicate that the information quality and satisfaction with the communication are rated lower during disruptions. Providing good and timely information during disruptions can be challenging due to the unpredictable nature of the situation. Participants reported a lack of information regarding alternatives and the cause of disruptions. Availability of such information could lead to a better understanding of the situation, improved patience, and decision-making. Additionally, information on facilities, such as coffee shops, restrooms, and Wi-Fi, was often rated lower in terms of quality and satisfaction and was frequently unavailable. Information about facilities could make a journey more pleasant and comfortable.

Limitations. While this research was conducted carefully, we recognise that there are some limitations. Firstly, the diary study was designed to comprehensively capture information needs and desires while limiting participants' burden. However, the setup limited our ability to fully comprehend the reasoning behind those needs and desires. Consequently, based on the current study, we can not completely understand why they have specific needs or desires in their travel situation. Moreover, the survey and diary study question setups could have been better aligned for comparison purposes, yet there is a decent overlap in responses and the findings from both methods can still complement each other.

Second, our research is solidly focussing on the public transport infrastructure in the Netherlands, which may limit generalizability to other countries. However, the differences in stated and revealed preferences remain applicable and are a valuable insight that should be considered in other contexts.

Lastly, most travels recorded in the diary study involved train journeys. This is no surprise since our sample was partially from the panel of the Dutch Railways, and the train is the most commonly used means of public transport in the Netherlands [9]. Yet, this may have skewed the input towards information needs/desires related to train travel.

Future Work. The associations found indicate opportunities to personalize travel information. The factors found to be associated with the different information types could be explored as input for personalized traveller information systems. The nature of this research is exploratory. Therefore, any identified associations should be interpreted merely as inspiration for further investigation to leverage these findings to model travellers' information requirements. Another example would be adapting information based on context, as done by [22].

Moreover, the results show the opportunity to improve the information provision during disruptions. Both the availability of certain information as well as the quality could be improved. Based on the results, the information types alternatives and disruption cause are important and have the potential to be improved upon. Facilities, even though they are not one of the most frequently mentioned information types, show the potential to improve the information provision for a certain traveller group.

Additional future work includes investigating information profiles in our sample [36]. This includes a classification based on the information requirements as well as the demographic and trip-related data.

6 Conclusion

This paper describes the outcomes of two studies on the information requirements of public transport commuters: a survey for stated information needs/desires and a diary study for revealed needs/desires. This resulted in a comprehensive list, which could be leveraged to further improve traveller information systems. Generally, during disruptions, information about the duration, cause of disruption, consequences for the journey, and alternatives are important. Information about potential disruptions, the schedule, journey planning, and interchanges are important for journeys without disruptions. The real-time information requirements differ from those reported from memory. Furthermore, a significant difference was found between the needs and desires during a disruption or a regular journey. Information provision could be improved during disruptions, for example, by addressing missing information about alternatives, the duration, or the cause of disruption. Moreover, tailoring information based on factors such as age group, travel frequency, station size, and disruption classifications presents opportunities for personalized public transport information systems.

Acknowledgements. We thank the participants for their time and effort. This work was partially funded by the Dutch Railways.

References

1. Adele, S., Tréfond-Alexandre, S., Dionisio, C., Hoyau, P.A.: Exploring the behavior of suburban train users in the event of disruptions. Transport. Res. F: Traffic Psychol. Behav. **65**, 344–362 (2019)
2. Armougum, A., Gaston-Bellegarde, A., Joie-La Marle, C., Piolino, P.: Physiological investigation of cognitive load in real-life train travelers during information processing. Appl. Ergon. **89**, 103180 (2020)
3. Bannert, M.: Managing cognitive load-recent trends in cognitive load theory. Learn. Instr. **12**(1), 139–146 (2002)
4. Bartram, D.: Comprehending spatial information: the relative efficiency of different methods of presenting information about bus routes. J. Appl. Psychol. **65**(1), 103 (1980)
5. Berggren, U., Brundell-Freij, K., Svensson, H., Wretstrand, A.: Effects from usage of pre-trip information and passenger scheduling strategies on waiting times in public transport: an empirical survey based on a dedicated smartphone application. Public Transport **13**, 503–531 (2021)
6. Brakewood, C., Watkins, K.: A literature review of the passenger benefits of real-time transit information. Transp. Rev. **39**(3), 327–356 (2019)
7. Cats, O., Koutsopoulos, H.N., Burghout, W., Toledo, T.: Effect of real-time transit information on dynamic path choice of passengers. Transp. Res. Rec. **2217**(1), 46–54 (2011)
8. Centraal Bureau voor de Statistiek: Hoeveel reisden inwoners van nederland van en naar het werk? (2021). https://www.cbs.nl/nl-nl/visualisaties/verkeer-en-vervoer/personen/van-en-naar-werk
9. Centraal Bureau voor de Statistiek: Hoeveel wordt er met het openbaar vervoer gereisd? (2021). https://www.cbs.nl/nl-nl/visualisaties/verkeer-en-vervoer/personen/openbaar-vervoer
10. Chen, L., Qi, L.: A diary study of understanding contextual information needs during leisure traveling. In: Proceedings of the Third Symposium on Information Interaction in Context, pp. 265–270 (2010)
11. Cochran, W.G.: The comparison of percentages in matched samples. Biometrika **37**(3/4), 256–266 (1950)
12. Costa, P.T., Jr., Terracciano, A., McCrae, R.R.: Gender differences in personality traits across cultures: robust and surprising findings. J. Pers. Soc. Psychol. **81**(2), 322 (2001)
13. Cramér, H.: Mathematical Methods of Statistics, vol. 26. Princeton University Press (1999)
14. Dunn, O.J.: Multiple comparisons among means. J. Am. Stat. Assoc. **56**(293), 52–64 (1961)
15. Farag, S., Lyons, G.: To use or not to use? an empirical study of pre-trip public transport information for business and leisure trips and comparison with car travel. Transp. Policy **20**, 82–92 (2012)
16. Government of the Netherlands: Sustainable public transport
17. Habib, M.A., Anik, M.A.H.: Impacts of covid-19 on transport modes and mobility behavior: analysis of public discourse in twitter. Transp. Res. Rec. (2021)

18. van Hagen, M., van Oort, N.: Improving railway passengers experience: two perspectives. J. Traffic Transp. Eng. **7**(3), 2328-2142 (2019)
19. Hörold, S., Mayas, C., Krömker, H.: Identifying the information needs of users in public transport. Adv. Hum. Aspects Road Rail Transp. **1**(2012), 331–340 (2012)
20. Huang, Z., Loo, B.P., Axhausen, K.W.: Travel behaviour changes under work-from-home (WFH) arrangements during covid-19. Travel Behav. Soc. **30**, 202–211 (2023)
21. Ibraeva, A., de Sousa, J.F.: Marketing of public transport and public transport information provision. Procedia Soc. Behav. Sci. **162**, 121–128 (2014)
22. Jevinger, Å., Johansson, E., Persson, J.A., Holmberg, J.: Context-aware travel support during unplanned public transport disturbances. In: VEHITS 2023–9th International Conference on Vehicle Technology and Intelligent Transport Systems, vol. 1, pp. 160–170. SCITEPRESS (2023)
23. Ku, D.G., Um, J.S., Byon, Y.J., Kim, J.Y., Lee, S.J.: Changes in passengers' travel behavior due to covid-19. Sustainability **13**(14), 7974 (2021)
24. Leng, N., Corman, F.: The role of information availability to passengers in public transport disruptions: an agent-based simulation approach. Transpo. Res. Part A Policy Pract. **133**, 214–236 (2020)
25. Mantel, N.: Chi-square tests with one degree of freedom; extensions of the mantel-haenszel procedure. J. Am. Stat. Assoc. **58**(303), 690–700 (1963)
26. McNemar, Q.: Note on the sampling error of the difference between correlated proportions or percentages. Psychometrika **12**(2), 153–157 (1947)
27. Mulley, C., Clifton, G.T., Balbontin, C., Ma, L.: Information for travelling: awareness and usage of the various sources of information available to PT users in NSW. Transp. Res. Part A Policy Pract. **101**, 111–132 (2017)
28. de Palma, A., Vosough, S., Liao, F.: An overview of effects of covid-19 on mobility and lifestyle: 18 months since the outbreak. Transp. Res. Part A Policy Pract. (2022)
29. Papangelis, K., Velaga, N.R., Ashmore, F., Sripada, S., Nelson, J.D., Beecroft, M.: Exploring the rural passenger experience, information needs and decision making during public transport disruption. Res. Transp. Bus. Manag. **18**, 57–69 (2016)
30. Pearson, K.: On the criterion that a given system of deviations given system of deviations of a correlated system of variables is such that it can be reasonably supposed to have arisen from random sampling. Philos. Mag. Ser. **5**, 157–175 (1900)
31. Pearson, K.: X. on the criterion that a given system of deviations from the probable in the case of a correlated system of variables is such that it can be reasonably supposed to have arisen from random sampling. London Edinburgh Dublin Philos. Mag. J. Sci. **50**(302), 157–175 (1900)
32. QSR International Pty.: Nvivo 20 (2023). https://www.qsrinternational.com/nvivo-qualitative-data-analysis-software/home
33. Romero, C., Zamorano, C., Monzón, A.: Exploring the role of public transport information sources on perceived service quality in suburban rail. Travel Behav. Soc. **33**, 100642 (2023)
34. Rosenthal, R., Cooper, H., Hedges, L., et al.: Parametric measures of effect size. In: The Handbook of Research Synthesis, vol. 621, no. 2, pp. 231–244 (1994)
35. Tang, L., Ho, C.Q., Hensher, D.A., Zhang, X.: Investigating traveller's overall information needs: What, when and how much is required by urban residents. Travel Behav. Soc. **28**, 155–169 (2022)
36. Un, P., Adelé, S., Vallet, F., Burkhardt, J.M.: How does my train line run? Elicitation of six information-seeking profiles of regular suburban train users. Sustainability **14**(5), 2665 (2022)

37. Van Lierop, D., Badami, M.G., El-Geneidy, A.M.: What influences satisfaction and loyalty in public transport? A review of the literature. Transp. Rev. **38**(1), 52–72 (2018)
38. Vianello, M., Schnabel, K., Sriram, N., Nosek, B.: Gender differences in implicit and explicit personality traits. Personality Individ. Differ. **55**(8), 994–999 (2013)
39. Wilcoxon, F.: Individual comparisons by ranking methods. In: Kotz, S., Johnson, N.L. (eds.) Breakthroughs in Statistics: Methodology and Distribution, pp. 196–202. Springer, New York (1992). https://doi.org/10.1007/978-1-4612-4380-9_16
40. Yeboah, G., Cottrill, C.D., Nelson, J.D., Corsar, D., Markovic, M., Edwards, P.: Understanding factors influencing public transport passengers' pre-travel information-seeking behaviour. Public Transport **11**, 135–158 (2019)

User-Centered Design of Land-Air Travel Service: HMI Design Strategies and Challenges

Yiqian Xu[1], Jieqi Yang[1], Ping Wang[1], Yuchen Wang[3], and Jianmin Wang[1,2,3](✉)

[1] Car Interaction Design Lab, College of Arts and Media, Tongji University, Shanghai 201804, China
wangjianmin@tongji.edu.cn

[2] Intelligent New Energy Vehicle Collaborative Innovation Center, Tongji University, Shanghai, China

[3] Shenzhen Research Institute, Sun Yat-Sen University, Shenzhen 518057, China

Abstract. Urban Air Mobility (UAM) contributes to alleviating ground traffic pressure, enhancing traffic efficiency, and promoting sustainable transportation development. The modular design of vehicles, exemplified by Airbus Pop.Up, further improves traffic efficiency. Currently, UAM development is in the preparatory stage, with a research gap in the study of cabin Human-Machine Interface (HMI). Based on the concepts of UAM and modular transportation, this study focuses on the cockpit HMI design for Land-Air Travel Service, referred to as LADUC (Land-Air Dual-Use Cabin) HMI. The research goal is to propose design guidance strategies for the usability and user-friendliness of HMI in LADUC.

In the preliminary phase, a combination of field research, desktop research, user studies integrating questionnaire surveys and in-depth interviews, along with Work Domain Analysis, was conducted. Based on the acquired user requirements and expectations, a functional architecture diagram for the central control screen and dashboard of LADUC was constructed. Usability tests were conducted in both low-fidelity and high-fidelity stages, with the test results indicating that the system has achieved preliminary usability.

Keywords: Urban air mobility (UAM) · User-centered design · HMI · Modular transportation · Work Domain Analysis (WDA) · Usability Test

1 Introduction

The Urban Air Mobility (UAM) Concept of Operations 2.0 [1], released by the Federal Aviation Administration (FAA) NextGen Office, defines UAM as a subset of Advanced Air Mobility (AAM) focusing on low-altitude airborne transportation in urban and suburban areas, utilizing highly automated Electric Vertical Takeoff and Landing (eVTOL) aircraft. Ground infrastructure to support eVTOL operations is known as a Vertiport, comprising elements like Vertipad (TLOF, FATO, Safety Area), charging stations, passenger platforms, towing carts, and related facilities [2].

H. Krömker (Ed.): HCII 2024, LNCS 14733, pp. 104–123, 2024.
https://doi.org/10.1007/978-3-031-60480-5_6

As a supplement to current urban transportation, UAM alleviates ground traffic congestion, enhances travel efficiency, and promotes sustainability. Schweiger, K., and Preis, L. [2], in their literature review, note a significant increase in UAM publications since 2016. According to Cohen, A. P. et al. [3], starting around 2010, companies like BLADE, SkyRyde, Uber, Oregon, and Airbus have progressively introduced air travel services through mobile applications (APPs). Notably, Airbus Voom reported 150,000 active APP users, 15,000 helicopter passengers, and a 45% customer retention rate from 2016 to 2020. Additionally, Morgan Stanley's 2018 Blue Paper projects the global urban air traffic industry revenue to reach $15 trillion by 2040 [4], indicating substantial market potential for UAM.

Currently, research in the field of UAM is predominantly concentrated in areas such as traffic management [5–7], ground infrastructure [2], user studies [8–10], legal policy analysis [11, 12], vehicle technology [13, 14], time efficiency [15], and cost analysis [16]. However, there exists a notable research gap in the Human-Machine Interface (HMI) within the airborne cabin to enhance passenger perceived safety and service acceptance.

Modular design is increasingly popular in transportation, involving breaking down vehicle components into modules that can be rearranged for different needs. Examples include MetroSnap, Airbus Pop.Up, and Scania's modular electric vehicle axle. Modular transportation brings efficiency benefits. A report by Roland Berger International Management Consulting indicates that Chinese cars have an average occupancy of less than 1.5 people, with about 95% idle time during the day. Separating the cabin from the motion module reduces idle time significantly, enhancing traffic efficiency. Therefore, there's a need for research on HMI for Land-Air Dual-Use Cabin (LADUC).

In UAM, the Pilot in Command (PIC) is the pilot onboard the aircraft or a remote pilot or an automated driving system who holds ultimate responsibility for flight operation [1]. While high-level UAM supports remote and automated driving, establishing a user-PIC relationship through HMI is crucial to address trust issues and boost safety perception and service acceptance in airborne autonomous or remote driving.

Therefore, this study focuses on the HMI for LADUC. The following research questions are proposed:

- How to design an HMI interface for LADUCs, ensuring information retrieval during flight is user-friendly and accessible?
- How to design communication and interaction interfaces with PIC to reduce user reaction time to hazardous conditions.

The research goal is to propose design guidelines for usable and user-friendly LADUC HMIs. In this research, a combination of field and desktop research is employed to categorize the functionality and content of airborne cabin HMIs. Preliminary user research is conducted through a combination of questionnaire surveys and in-depth interviews. Conducting prototype usability testing involves the integration of methods such as Wizard of Oz and real-time probes.

The research contributions include:

- Filling the research gap in LADUC HMI, exploring and categorizing the functional information architecture of airborne central screen and dashboard.

- Exploring LADUC HMI design to achieve usability and user-friendliness and proposing relevant design guideline strategies.

2 Related Work

2.1 UAM Research Progress

Due to the UAM development being in its preparatory phase, current research predominantly centers on legal policies, technological support, traffic management, and benefit assessments, with limited attention to the HMI domain. This section will primarily delve into the progress within the field of UAM human-machine interaction.

A crucial aspect of UAM HMI involves managing the relationships among drivers, passengers, and autonomous systems. NASA's UCAT proposes the UAM Maturity Levels (UML) model, categorizing UAM development into initial, transition, and mature stages. Mature stages support vehicle automation and remote management, ultimately aiming for personalized transport to meet societal expectations [17].

In the FAA CONOP [1], the level of aircraft automation is linked to the Pilot in Command's (PIC) engagement, classified into three levels: Human-Within-the-Loop (HWTL), Human-on-the-Loop (HOTL), and Human-Over-the-Loop (HOVLP). Analyzing different automation levels, Vempati et al. [18] utilized the Decision Identification Framework (DIF) and Human Automation System Oversight (HASO) Framework to identify key system design features influencing driver operations. In the HWTL stage, they advocate providing automation support to drivers in non-critical situations, executing repetitive routine operations, and presenting decision-support information to help maintain situational awareness and facilitate mutual understanding between humans and machines. In the HOTL stage, the focus is on keeping the driver in the loop, requiring the provision of necessary information perceived by the driver without presenting detailed information. In the HOVLP stage, consistent, transparent, and unbiased decision support information should be presented to support the passive monitoring of autonomous driving by the driver, maintaining situational awareness.

Pongsakornsathien et al. [19] employed an Applied Cognitive Task Analysis based on the OODA (Observe, Orient, Decide, and Action) model to establish HMI visual information and airspace planning requirements for Air Traffic Controllers (ATCo) and Unmanned Aircraft Systems Traffic Management (UTM) under Sheridan Level 5 and Level 7 scenarios. Their focus lies in the interpretability of AI and support for situational awareness among operational personnel.

User research serves as a crucial foundation for HMI, offering insights into user needs, challenges, and expectations. In the context of UAM, public acceptance is identified as a critical prerequisite for its developmental stages [20].

Yavas and Tez [10] constructed the UAM Acceptance and Usage Model (UAM-AUM), measuring seven dimensions, including intention to fly (IF), UAM affordability (UA), UAM conceptual intention (UCI), environmental consciousness (EC), general reliability (GR), perceived usefulness (PU), and behavioral intention to use (BIU). Through structural equation modeling (SEM), it was highlighted that PU is strongly correlated with BIU and acts as a mediating regulator for the impact of IF-UA-UCI-EC-GR on BIU. The key factor fostering UAM system acceptance is GR, linked to users' perception of uncertainty regarding the safety of the UAM system, encompassing aspects like emergency service applications and all safety-related elements (e.g., seatbelts, air quality, cabin indicators).

Tepylo et al. [21] provided a comprehensive overview of public acceptance, considering technological aspects like safety, sustainability, access equality, low noise, multimodal connectivity, local workforce development, and purpose-driven data sharing. Individual users prioritize factors such as cost, service provider reputation, safety, environmental impact, time savings, route planning, past experiences, and personal preferences. Enterprise users focus on system alignment with operational needs, while government institutions prioritize operational capabilities in scenarios like Remotely Piloted Aircraft Systems (RPLS) for monitoring and search and rescue tasks.

In a study on Personal Aerial Vehicles (PAVs), factors like weather conditions, battery status, time to destination, route selection, traffic information, driving speed, conditions of nearby traffic, speed limits, PAV images, and surrounding bird activity were highlighted as essential for in-cabin screen displays [22].

2.2 Pilot-in-Command

The role of the Pilot-in-Command (PIC) is vital for ensuring user travel safety. Advances in autonomous, sensing, and communication technologies have made the collaborative control of N aircraft by m operators (m:N, N > m) feasible. In 2005, the FAA issued the first airworthiness certificate for remotely piloted aircraft (UAV) [23]. Tepylo et al. [21] categorized remotely piloted aircraft into four main types: nano category, off-the-shelf category, passenger-carrying capable category, and the beyond category. The representative of the passenger-carrying capable category includes UAM aircraft and traditional Medium Altitude Long Endurance (MALE) Remotely Piloted Aircraft Systems (RPLS), currently operated by commercial or government entities but potentially evolving towards PAVs.

In the m: N remotely piloted aircraft control mode, the operator (akin to the PIC) interface requires in-depth investigation. Chandarana et al. [24] evaluated operator capabilities in handling sudden events and completing vehicle handovers using the Vigilant Spirit Control Station (VSCS) software. Results showed a higher frequency of user engagement in handovers under assisted handover with a greater number of sudden events, indicating a preference for assisted handover.

In summary, technologies related to UAM aircraft PIC are still in development. Given the strong correlation between PIC and travel safety, impacting user acceptance and trust, further research in PIC domain within the context of LADUC is crucial.

3 Methods

This study concentrates on crafting a user-friendly HMI for LADUCs in the context of UAM. Initially, a combination of field and desktop investigations was conducted, comparing HMI interfaces in commercial planes and helicopters. Secondly, through surveys and interviews, user opinions on the novel land-air travel service and preferences for LADUC HMIs were gathered, aligning with a user-centric design approach. Besides, guided by Work Domain Analysis (WDA), traditional airborne cabin HMI features were refined for the LADUC context, leading to the development of the layout and functional architecture of the LADUC HMI. Next, a low-fidelity prototype underwent initial usability testing, driving adjustments for a high-fidelity prototype. The second round of usability test based on high-fidelity prototype, using methods like "Wizard of Oz" and "Real-Time Probing," produced design guidelines for the LADUC HMI.

3.1 Field Research and Desktop Research

The purpose of field and desktop research is to summarize and generalize the general functionality and design of in-cabin HMI for aircraft. Despite differences in flight principles and driving modes among various aircraft types, there is a notable similarity in the core functions and content displayed on in-cabin HMI interfaces. Researchers experienced a simulated Airbus A320 flight at the CityFlight Shanghai Flight Experience Center in Shanghai, China, to observe and study HMI interface features. Additionally, for helicopter cabin HMI interfaces, researchers conducted desktop research using online videos, focusing on the Robinson R44 helicopter as a representative example.

Comparing the HMI interfaces of the Airbus A320 and Robinson R44 cockpits, along with other desktop research findings, six major functional modules for in-cabin HMI were identified. These modules encompass Vehicle Monitoring, Navigation, Communication, Auto/Assist Driving, Safety, and Adjustment Settings.

1. **Vehicle Monitoring:** Sub-functions include attitude indicator, altimeter, Variometer, airspeed indicator, heading indicator, pressure gauge, vaporizer temperature gauge, etc.
2. **Navigation:** Primarily involves GPS with aircraft icon, track, speed vector display, etc.
3. **Communication:** Sub-functions encompass interphone system, communication radio, transponder, etc.
4. **Auto/Assist Driving:** Sub-functions include speed control, heading control, altitude control, climb altitude control, etc.
5. **Safety:** Sub-functions comprise various types of warning messages (e.g., short-term conflict warning, danger zone intrusion warning, low altitude warning, etc.), different types of warning lights (e.g., carbon monoxide warning light, low fuel level warning light, etc.), critical instrument backup (e.g., additional mechanical backup instruments, the same display screen showing different instrument interfaces via buttons), main and reserve tank settings, inadvertent touch protection (e.g., pressing a button inward to activate autopilot mode, pulling a button outward to deactivate autopilot mode), emergency locator transmitter switch, etc.
6. **Adjustment Settings:** Sub-functions include cabin heating switch, cabin ventilation switch, dashboard backlight adjustment knob, etc.

3.2 User Research

The objective of user research is to explore, from the user's perspective, expectations and pain points regarding the new Land-Air travel service and attitudes towards the concept of PIC. The goal is to identify opportunities for designing an effective LADUC HMI interface. The research comprises both questionnaire surveys and in-depth interviews, with the latter building upon the insights gained from the former. The user research focuses on five key sub-questions:

Q1: What is the users' initial impression of the new Land-Air travel service?
Q2: Which travel stages impact users' perceived safety during the journey? Why?
Q3: What information provided by HMI can enhance users' perceived safety?
Q4: What are users' service expectations regarding the concept of a PIC?
Q5: How high is the users' acceptance of manual driving service modes?

Survey Structure. Before officially starting the research, it is crucial to clarify the purpose of the study to users. The aim is to understand consumer expectations, initial acceptance, and concerns regarding the new Land-Air travel service. Additionally, users should be introduced to the basic concept of the service, covering its definition, service process, and features.

The Land-Air travel service refers to an integrated aerial taxi system combining ground and airborne travel. The eight-stage service process encompasses departure, ground travel to the station, mode switching, takeoff, landing, mode switching, ground departure, and final arrival (Fig. 1). Aligned with UAM objectives [4, 5], the service exhibits characteristics including safety exceeding that of driving, heightened efficiency in short-distance travel, a cost estimate of 10–12 yuan per kilometer, constrained airborne travel routes, and highly autonomous land-air vehicles devoid of an in-cabin pilot. Operational modes comprise autonomous, manual (subject to necessary credentials), or remote pilot operation.

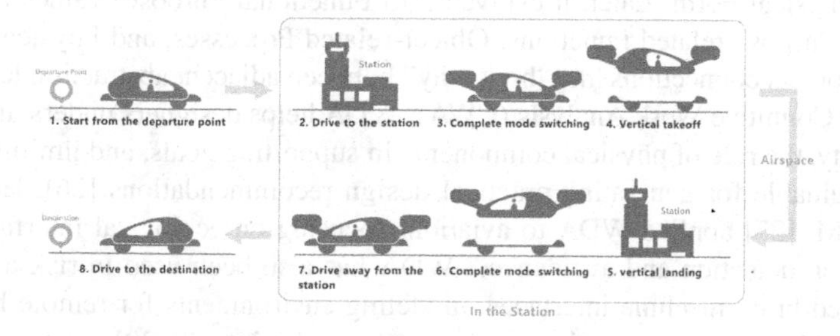

Fig. 1. Land-Air service process

The survey consists of 15 questions divided into five sections. Initially, participants provide demographic details such as gender and age (q1–q2). Following that, they share their first impressions of the land-air travel service, prioritizing concerns related to safety, pricing, travel efficiency, and service experience, along with scenarios and reasons for choosing this service (q3–q5). The third section delves into travel stages significantly affecting users' safety perceptions, ranking unsettling stages and influencing factors during takeoff/landing, mode transition, and airborne travel (q6–q10). Furthermore, participants express expectations regarding PICs, including preferred relationship modes and desired features of the PIC's cabin HMI (q11–13). Lastly, the survey explores users' acceptance of manual driving services, assessing the impact on safety perceptions and willingness to invest time in obtaining a flying license (q14–q15).

After the survey, six participants volunteered for in-depth interviews, conducted one-on-one online in a semi-structured format and recorded with consent. A standardized textual transformation was applied to the recorded oral narratives post-interview. The interview, featuring seven key questions and allowing flexibility, began with a brief introduction to the land-air travel service assuming participants had read the survey instructions.

The seven questions aimed to explore: ① reasons for choosing land-air travel in specific scenarios; ② the ideal land-air travel service process from users' perspective; ③ reasons for feeling unsafe during different travel stages; ④ HMI information needs to enhance perceived safety; ⑤ user acceptance of obtaining a flying license; ⑥ expectations regarding PICs; and ⑦ user expectations and concerns regarding land-air travel services.

Participants. The survey yielded a total of 56 valid responses. Regarding demographic information, 17 male and 39 female participants completed the questionnaire. The majority of users (55.36%) fell within the 18–25 age range, potentially representing a target user group for future land-air travel services or UAM. Eight participants volunteered for in-depth interviews, consisting of 2 males and 6 females (average age 24.88).

3.3 Work Domain Analysis (WDA)

Work Domain Analysis (WDA), introduced by Rasmussen, is a method for designing interactions in complex systems [25]. WDA consists of five abstraction hierarchy (AH) levels: Functional Purpose, Abstract Function, Generalized Function, Physical Function, and Physical Form. Later, it evolved into Functional Purpose, Values & Priority Functions, Purpose-related Functions, Object-related Processes, and Physical Objects. There are logical connections like "how-why" between adjacent abstraction levels. As a key part of Cognitive Work Analysis (CWA), WDA helps designers understand system functionality, the role of physical components in supporting goals, and limiting factors. WDA is valuable for generating practical design recommendations [26]. Ho, D. and Burns, C. M. [27] applied WDA to aviation, focusing on ecological interface design for "collision detection and avoidance." WDA has also been used in risk assessment for advanced brain-machine interfaces, modeling environments for remote healthcare services, and human factors analysis in autonomous driving [28–30].

In this study, WDA is employed to outline the main functions of the HMI architecture for the LADUC, with subsequent research refining dashboard and central control screen functionality.

3.4 Usability Test Based on Low-Fidelity Prototype

Summarizing the first three research phases, we focused on the central control screen and dashboard for design. We started by creating a functional architecture diagram and then developed a low-fidelity prototype. To validate design alignment with user mental models, usability testing is essential at each design stage. This testing primarily evaluates efficiency (task completion time), effectiveness (successful task completion), and user satisfaction [31].

This study's usability testing for the low-fidelity prototype mainly aims to assess interface effectiveness and efficiency. Participants will use the early interface design, and feedback and performance data will be collected to identify potential design flaws early on, guiding future iterations.

Participants. Three participants with driving experience took part in the usability testing of the low-fidelity prototype, with an average age of 23 years.

Apparatus. The experiment setup involved computer-based low-fidelity prototypes for the central control screen and dashboard, a control column, a paper-based warning screen prototype (simulating the helicopter cockpit's warning light system), an actor as the PIC and interview recording equipment and cameras.

Experiment Design. The experiment employed "Wizard of Oz" (WoZ) and real-time probing as testing methods. It consisted of two tasks: a normal driving scenario and an emergency scenario. In the normal task, participants interacted with the central control screen prototype in an automated cockpit simulator to perform basic cockpit settings. In the emergency task, participants simulated responding to the PIC's commands in an emergency, involving interactions with the central control screen prototype and control column to ensure safety.

- Normal Task 1: Participants freely explore and share opinions on the interface prototype aloud.
- Normal Task 2: Locate the interface for accessing entertainment information.
- Normal Task 3: Participants set ground-side motion status during automated flight.
- Emergency Task: Respond promptly to the danger alarm, find the "PIC contact" icon, follow instructions to navigate the aircraft safely, and then end the PIC's call.

Procedure. First, place the control column, warning screen paper prototype, and low-fidelity prototypes of the central control screen and dashboard in their designated positions. Then, invite three participants to individually conduct the experiments. Prior to starting, provide participants with basic training and familiarize them with the central screen and control column. After completing all tasks, conduct interviews with the participants. Record the entire experiment, including screen recordings of participants' operations on the central control screen.

3.5 Usability Test Based on High-Fidelity Prototype

Participants. Six participants with driving experience took part in the usability testing of the high-fidelity prototype, with an average age of 23 years.

Apparatus. Similar to the low-fidelity prototype usability testing, this testing will be conducted in a quieter and less distractive environment (Fig. 2).

Experiment Design. The high-fidelity prototype aims to confirm the effectiveness of iterative enhancements. Participants interact with the updated interface to demonstrate its usability. Results from questionnaires and research guide future iterations. The experiment includes usability test for both the dashboard and central control screen.

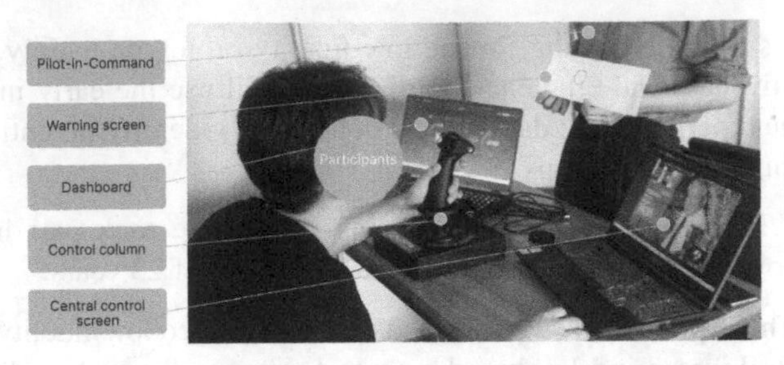

Fig. 2. Usability Test Experimental Equipment and Environment

Usability test for the dashboard requires participants to observe and describe the meaning of parameters displayed on the left and right dials, along with the current motion status and other interpretable data on the interface. Test for the central control screen includes both normal and emergency tasks.

Compared to the low-fidelity prototype usability test, only the normal task 1 has been modified to "Request the participants to observe the navigation interface on the central control screen. First, read relevant information about the current journey, including the starting point, time, remaining distance, etc. Second, gather information about the surroundings, such as nearby stations, alternate landing points, etc." The other normal tasks and emergency tasks remain unchanged.

Procedure. The procedure is generally similar to the low-fidelity prototype usability test. However, after completing all tasks, participants will fill out a system and functionality usability scale. The functionality usability scale includes questions addressing the deficiencies identified in the low-fidelity prototype to validate the effectiveness of the iterations.

4 Results

4.1 User Research

Integrating the results from the questionnaire survey and in-depth interviews, the findings for the five research sub-questions are summarized as follows:

First Impressions. Users' first impressions of Land-Air travel services reveal their initial attitudes and predispositions before experiencing the actual service. Understanding these impressions helps optimize the user experience through HMI design, focusing on addressing user concerns and enhancing their expectations. This approach aims to enhance users' expectations and mitigate or address their concerns, thereby shaping their overall attitude towards the service.

Consumer-Driven Factors. Questionnaire q3 assessed participant priorities regarding safety, experience, price, and efficiency for Land-Air travel services. Results indicate safety as the top concern, despite the premise of services being twice as safe as driving. Price and travel efficiency closely follow, with service experience ranking the lowest.

This implies that users will prioritize safety, price, and travel efficiency when considering land-air travel services. Therefore, the marketing strategy should highlight high safety standards, competitive pricing, and efficient travel. Service experience, influenced by factors like perceived safety and actual costs, determines users' decisions to use these services again. Consequently, in designing cabin HMI for a significant impact on the travel experience, the focus should be on enhancing perceived safety, emphasizing affordability, and highlighting efficiency improvements (Table 1).

Table 1. Ranking Results of User Concerns for Safety, Experience, Price, and Efficiency

Property	Score	1st Place	2nd Place	3rd Place	4th Place	Subtotal
Safety	**3.39**	35(62.50%)	12(21.43%)	5(8.93%)	4(7.14%)	56
Price	**2.52**	10(17.86%)	17(30.36%)	21(37.50%)	8(14.29%)	56
Efficiency	**2.45**	6(10.71%)	23(41.07%)	17(30.36%)	10(17.86%)	56
Experience	**1.64**	5(8.93%)	4(7.14%)	13(23.21%)	34(60.71%)	56

Travel Purpose. Based on the survey results from q4, approximately 78.57% of users prefer Land-Air travel for trips between nearby cities, 58.93% for travel between the city and suburbs, and 28.57% for Inter-suburban travel. In response to q5, it was found that users primarily opt for land-air travel for tourism (58.93%), business trips (62.50%), and airport journeys (57.14%). Commuting to work represents a lower preference at 33.93%. Further analysis reveals that pricing and travel efficiency are key factors influencing user choices.

Cross-analysis of q4 and q5 (Fig. 3) indicates that business trips are the most common purpose for users choosing Land-Air travel between nearby cities. Inter-suburban travel is the scenario with the highest likelihood of users choosing Land-Air travel services for commuting to work. When users choose land-air travel for travel between the city and suburbs, going to the airport is the most common travel purpose.

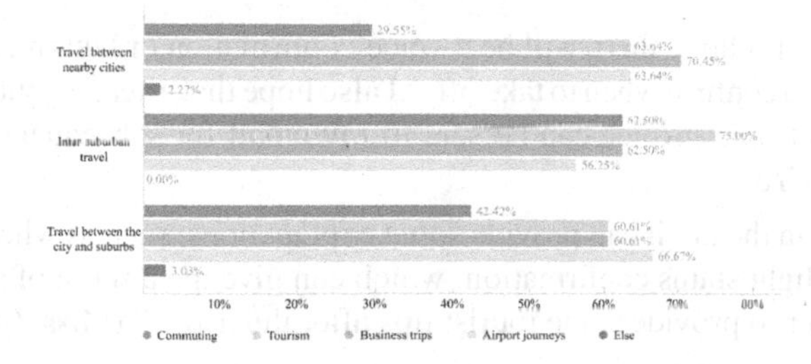

Fig. 3. The cross-analysis results of questionnaires q4 and q5

The analysis indicates that "convenience" and "travel efficiency" are crucial keywords for travel purposes, which are further supported by in-depth interviews.

In the first interview question, participants cited reasons for choosing land-air travel services in specific scenarios. Key themes included "travel efficiency," "time urgency," and "traffic congestion," mentioned a total of 5 times. Additionally, phrases like "inconvenience" were mentioned twice.

> "I might use map software to confirm traffic congestion in advance, and if there is traffic congestion, I would prefer to use this service…"(Mrs. Xu)
>
> "Since take-off and landing will also involve the information dispatching of the tower, which will occupy time, so under all perfect systems, it may be chosen only when there is a long commute and the road is congested…"(Mr. Mei)

This indicates that users choose Land-Air travel services primarily due to congestion in ground transportation, extended travel times, and high time pressure. Therefore, when designing the cabin HMI, it is crucial to emphasize the service's superior travel efficiency. This can be achieved by displaying real-time "estimated time of arrival" and notifying users during order-related activities about the time saved, such as "This trip will save you XX (time)."

> "I would choose to use it in short-distance travel, because it can avoid the problem of being crowded with people on public transportation. You know, it's very troublesome to carry luggage and transfer on public transportation during travel…"(Miss Zou)

To improve the user travel experience, the cabin HMI design should consider users' preference for Land-air travel services in suburban areas with multiple transfers and a desire to avoid crowded situations during tourism. This can be achieved by incorporating features such as an entertainment information system and adaptive cabin settings linked to user IDs to record seat preferences.

Expectations & Concerns. During the second part of the in-depth interview, users are asked to provide additional details about their ideal Land-Air travel service using a timeline as a guide. The Laddering interview method is used to uncover the reasons behind specific service preferences, aiming to identify opportunities for LADUC HMI design.

> "Before take-off, I hope there will be a voice confirmation or button confirmation for customers to confirm when to take off…I also hope that after the journey is completed, an electronic invoice can be sent to my email for subsequent reimbursement…"(Miss Yu)
>
> "While flying in the air, it can provide some explanations, such as where the flight is located or flight status confirmation, which can give me a sense of security…It would be better to provide some tourist tips after the trip…" (Miss Zou)

In-depth interviews provided user insights into specific service details, focusing on takeoff, flight, mode switching, and the post-trip phase. For takeoff, users emphasized the need for user confirmation. In-flight, their core requirements included small tables, power outlets, and entertainment options. They also suggested location narration and flight status confirmation for enhanced security. During mode switching, improving safety perception was highlighted. Post-trip, users sought quick payment methods and a convenient invoice issuance process.

In response to the question about expectations and concerns, five users expressed concerns about "travel efficiency," and safety-related terms were mentioned seven times. Other concerns encompassed station locations, noise, cabin cleanliness, and the difficulty of learning flying skills.

In conclusion, the LADUC HMI design should prioritize enhancing perceived safety, emphasizing travel efficiency advantages, and improving the overall travel experience (Table 2).

Perceived Safety. Before experiencing Land-Air travel services, users' perceived safety during various travel stages may differ due to personal factors. Questionnaire q6 reveals two trends:

- Takeoff/landing and mode transitions are more likely to cause insecurity, especially during takeoff/landing.
- As users approach their destination, there is a tendency for an increase in the perceived sense of security.

Table 2. Ranking results of user insecurity at different stages of travel

Stage	Score	1st Place	2nd Place	3rd Place	4th Place	5th Place	6th Place	7th Place	8th Place	Subtotal
4	5.7	14	13	7	12	1	2	1	2	52
3	5.46	14	10	11	8	4	0	1	0	48
5	5.2	10	13	6	9	7	2	2	1	50
6	4.95	6	5	17	10	1	12	1	0	52
1	3.46	7	3	4	0	14	6	7	5	46
2	3.18	2	6	4	3	8	11	7	2	43
7	2.36	2	2	1	5	5	4	17	5	41
8	1.59	1	1	2	1	3	4	4	25	41

Additionally, q7–9 explored factors contributing to insecurity during takeoff/landing, mode switching, and airborne flight. For a comprehensive understanding, each stage considered technical safety, safety checks, driving control, and the absence of the pilot as potential causes, with specific details varying. This detailed breakdown offers precise guidance for subsequent HMI design. Results showed:

- Takeoff/landing: Safety checks (uncertainty about personal readiness) > Technical safety > Driving control > Absence of the pilot;
- Mode switching: Safety checks (uncertainty about mechanical stability) > Technical safety > Driving control > Absence of the pilot;
- Airborne flight: Technical safety > Driving control > Safety checks (uncertainty about real-time operation) > Absence of the pilot.

Analysis shows that users are highly accepting of autonomous and remotely monitored driving. Compared to other factors, the absence of the driver has a relatively low impact on users' sense of travel safety. During takeoff/landing and mode switching stages, safety checks have the most significant impact on users' perceived insecurity, indicating a need for information about phase transitions and vehicle operating status. In the airborne flight stage, driver control has a notable impact on user safety, confirming the influence of flight driving knowledge or skills on users' sense of security.

The fourth interview question delves deeper into understanding the causes of user insecurity during various travel stages through open-ended inquiries:

> "The sudden feeling of weightlessness during take-off and landing may make me feel unsafe. I feel that both mode switching and take-off/landing need real-time re-minders..." (Mrs. Ni)
>
> "When switching modes, I may feel nervous and anxious because I'm not sure if the hardware switching information is visible..." (Mr. Zhen)

During interviews, users expressed unease during takeoff/landing due to factors like weightlessness and shaking. In the mode-switching phase, concerns about mechanical stability were raised, and during airborne travel, issues such as shaking, collision risks, and survival rates in accidents contribute to insecurity.

Design strategies based on survey and interview findings include:

- Improving transparency on stage transitions and vehicle status.
- Providing statistics, technical explanation videos, and emergency guidelines to endorse safety.
- If manual driving modes are offered, real-time feedback on driving inputs is essential.

Concerning physical vs. virtual buttons, q10 showed that the emergency alert button's interaction mode does not significantly affect user safety perception. However, over half of the users prefer a physical button, considering its critical role in emergencies for a more direct and reliable safety mechanism.

HMI Function Related. The in-depth interviews addressed the HMI information that enhances user safety. Based on the general cockpit HMI functional modules, the summarized aspects are as follows:

- Vehicle Monitoring: Highlights battery status, flight altitude, and vehicle operational status (highlighting safety-related information), along with maintenance details, cockpit hardware checks, and process display for mode switching.
- Navigation: Provides travel information like distance, time, and route to the destination, along with weather updates.
- Communication: Allows seamless AI and human customer service transitions.
- Automatic/Assisted Driving: Presents driving modes such as automatic, manual, and remote driving.
- Safety: Indicates emergency survival equipment location, safety check results after mode switching, and prominent emergency assistance buttons.

Pilot-in-Command Related. The PIC's role in ensuring safety during airborne phases was investigated in q11 to 13. q11 explored users' preferences for their relationship with the PIC, affecting their virtual representation and interaction mode. Users favored real-time communication with the driver across all age groups, with a stronger preference among older users. Younger users showed higher acceptance of a consumer-service relationship, while a remote driver-passenger relationship was generally not preferred.

While real-time communication during the service promotion phase is feasible, its long-term implementation cost is high. In-depth interviews were conducted to understand users' preferences for specific relationships, guiding HMI design. The interviews revealed that acceptance of a consumer-service relationship is influenced by factors such as service universality, technological maturity, flight driving experience, service usage experience, and the demand for personal travel space.

Questionnaire q12 explored if the continuous display mode of the digital PIC would enhance users' perceived safety, with nearly all users responding positively. Questionnaire q13 investigated the preferred form and display location of the digital remote pilot. Results showed that users favor holographic projection as the display mode, followed by the digital image on the central control screen, the digital image on the dashboard, and the in-car robot on the table. Users' preference for holographic projection may be due to its immersive communication style. In practical applications, a combination of multiple display methods can be considered.

Summarized HMI design guidelines for PIC:

- In the early service development stages, consider providing users with low-latency real-time contact methods.
- To control service costs, consider using a virtual PIC for real-time companionship and Q&A services. Additionally, to improve service timeliness and effectiveness, consider employing a human PIC for prompt feedback after a call.

Manual Driving Acceptance. To examine how driving control affects users' perceived safety during airborne flights and assess users' willingness to obtain a flying license, we conducted surveys (q14–15). Results show that about three-quarters of users believe having a flying license enhances travel safety, and they are willing to invest time and effort in obtaining one. In-depth interviews revealed users' willingness to get a flying license under conditions of lower costs, higher service popularity, and increased usage demand. Simplifying the complex vehicle monitoring functions in the general flying cabin's driving control interface and using knowledge transfer methods can reduce the learning curve for users, ensuring usability.

4.2 Work Domain Analysis (WDA)

WDA in this study focuses on "enhancing travel safety," with the specific content of AH illustrated in Fig. 4. Given the study's focus on HMI elements, particular attention is directed towards Object-related Processes connected to the interactive interface in Physical Objects (highlighted in gray), including "Presentation of driving status," "Presentation of status of other traffic participants," "Hazard warning," "Call operation," "Remote indication," and "Emergency Operation Guidelines."

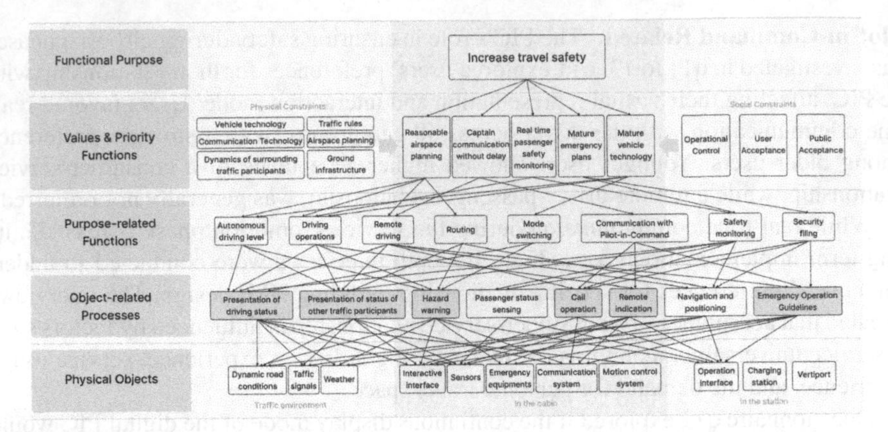

Fig. 4. The Abstraction Hierarchy (AH) work domain model of Land-Air travel service

4.3 Usability Test Based on Low-Fidelity Prototype

Figure 5 depicts the functional architecture diagram for the central control screen and dash, while Fig. 6 illustrates the low-fidelity prototype of the central control screen.

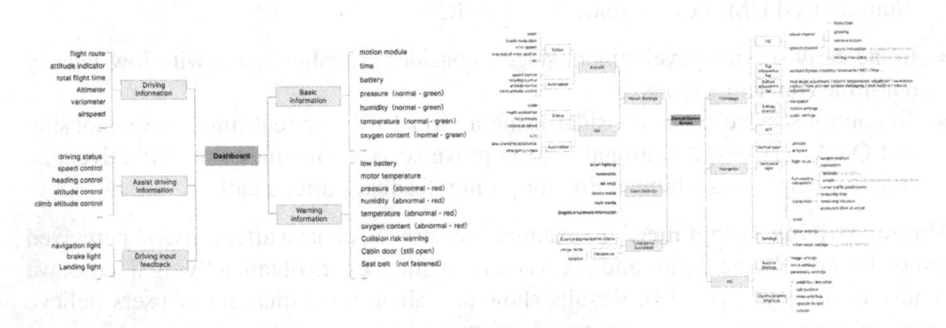

Fig. 5. Functional architecture diagram of central control screen and dashboard

Objective Data Analysis. All users in the usability test completed tasks successfully, confirming the effectiveness of the central control screen HMI design.

Post-test analysis of user interactions revealed confusion regarding entry points for Motion Settings and Cabin Settings. Users also faced difficulties accessing vehicle settings directly from the homepage. To enhance semantic clarity, it is recommended to modify the entrance icons on the homepage. For emergency tasks, participants hesitated briefly before selecting the call button and encountered challenges finding the "end call" button. Design adjustments are suggested to clearly distinguish the emergency call button from routine inquiry buttons.

Video analysis showed that during emergency tasks, participants actively communicated with the PIC and frequently shifted their gaze between the instructions on the warning screen and the remote pilot. This indicates the effectiveness of presenting driving operation steps on the warning screen.

Subjective Data Analysis. Participants provided feedback and suggestions for the interface. For normal tasks, they recommended improvements in font size, color contrast, and icon semantics. They also suggested an independent entry for the entertainment module alongside the three setting modules. In emergency situations, participants wanted clearer indications on the alarm button, including changes in size, shape, and color, to attract the driver's attention proactively.

4.4 Usability Test Based on High-Fidelity Prototype

The design philosophy for the LADUC central control screen includes: ① Considering journey information and safety status as the most crucial for users, the homepage features a one-click switch between the navigation and motion status pages using the "Switch Views" button; ② Three entrances on the homepage—Navigation, Motion Settings, and Cabin Settings—where Motion Settings display the current settings for the connected motion module; ③ A direct call entrance for the PIC is at the bottom of the homepage, allowing users to connect with a live PIC or an AI PIC; ④ The PIC calling interface includes command records and speech-to-text functionality for user convenience; ⑤ The bottom status bar offers quick settings for seats and air conditioning, while Cabin Setting provides detailed adjustments.

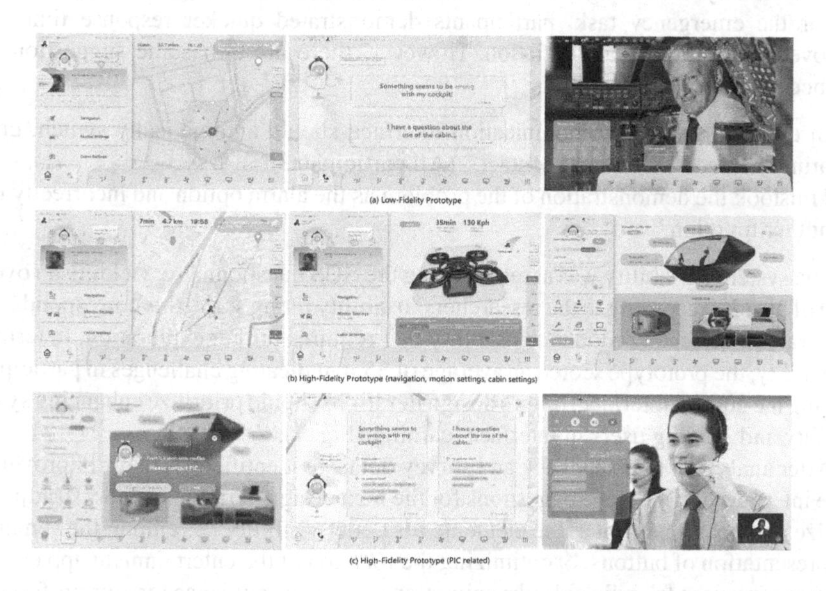

Fig. 6. Partial low-fidelity and high-fidelity prototype of central control screen (The PIC calling interface and user avatars utilize online images, while the remaining components are original creations.)

Subjective Data Analysis. In the dashboard part, approximately two-thirds of users hesitated, made errors, or missed information. Recognition issues occurred with icons related to music playback and backup power, especially among novice users struggling with autonomous driving parameters and aircraft status indicators. Feedback included:

- "In representing altitude levels, the use of a dial is not appropriate."
- "The bird's-eye view requires a high level of spatial imagination for route display."

During the central control screen usability test, participants provided several recommendations. For the completion of normal task 1, the following suggestions were received:

- "I wish the display of altitude levels could be interactive, allowing me to set a target altitude."
- "Some details on the interface are too small, making them difficult to discern."
- "The icons for surrounding information are somewhat abstract."

During the completion of normal task 2, the following recommendations were received:

- "I need a global automation setting, and detailed settings when manual driving."
- "I suggest adding explanations to assist my understanding of the settings."

During normal task 3, half of the participants found the entrance to the entertainment app successfully, indicating a need for further optimization in its positioning.

For the emergency task, participants demonstrated quicker response times and improved accuracy in call initiation. However, there are still some suggestions for enhancement, including:

- "In emergencies, the communication interface should automatically switch, eliminating the need for manual contact." (All participants)
- "I mistook the demonstration of the problem as the alarm option and incorrectly used another function."

The system's usability was assessed using the SUS questionnaire, yielding an overall score of 59.17. According to the researchers' usability rating scale, this corresponds to an "OK" rating, indicating moderate usability that requires further exploration. In terms of learnability, the prototype scored an average of 35.4, indicating challenges in participants learning the current interface. Future design iterations should prioritize enhancing system usability and guiding users in interface learning (Fig. 7).

After analyzing user feedback and survey data, we identified areas for improvement in the interface design. Key suggestions for the next design iteration include: ① adjusting the size and position of interface elements for better visibility, ② refining the semantics and presentation of buttons, ③ optimizing the location of the entertainment app entry, ④ incorporating user-friendly aids like animations and voice guidance for non-professional users, and ⑤ implementing a graded approach for handling emergency situations (in high urgency scenarios, the PIC can directly intervene, triggering automatic call initiation and alert mode to minimize user stress or errors. In low urgency situations, users can decide

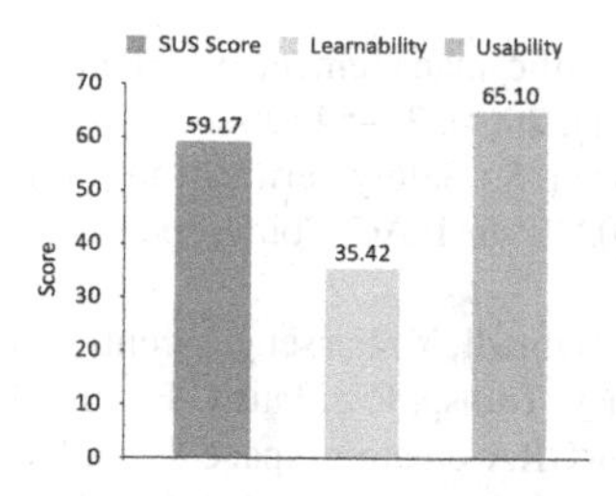

Fig. 7. Average data of SUS system availability scale

whether to contact the PIC themselves.). These refinements aim to enhance the overall usability and user experience.

5 Discussion and Conclusion

Based on preliminary research, LADUC HMI design objective is to enhance user perceived safety, travel efficiency and the overall travel experience. Guided by WDA, identified HMI features encompass motion status, navigation information, hazard warnings, PIC calls, remote commands, and emergency operation guidance.

Usability testing, progressing from low-fidelity to high-fidelity prototypes, indicated that the current HMI meets usability standards but requires further optimization. A summary of optimization directions has been outlined for future research.

It's important to note that this study focused on trial users from major Chinese cities, introducing geographical and cultural limitations. The relatively small sample size during user research and usability test may impact result credibility. Nonetheless, these findings can serve as a reference for LADUC HMI design. Additionally, the integration of land-air travel services as a supplementary transportation mode can be explored. In dense urban traffic scenarios, the land travel phase within the land-air travel service may experience reduced efficiency due to traffic flow. In such cases, the associated app can recommend users to use other transportation modes to reach nearby stations before initiating the airborne journey, reducing costs and enhancing efficiency.

Acknowledgments. This research was supported by the Fundamental Research Funds for the Central Universities; the High-Level Foreign Expert Project (G2022133031L, G2023133041L); Shenzhen Science and Technology Program (GJHZ20220913142401002).

References

1. Urban Air Mobility (UAM) Concept of Operations. Federal Aviation Administration (2023)
2. Schweiger, K., Preis, L.: Urban air mobility: systematic review of scientific publications and regulations for vertiport design and operations. DRONES **6**(7) (2022)
3. Cohen, A.P., Shaheen, S.A., Farrar, E.M.: Urban air mobility: history, ecosystem, market potential, and challenges. IEEE Trans. Intell. Transp. Syst. **22**(9), 6074–6087 (2021)
4. EHANG, Future Mobility: White Paper on Urban Air Transport Systems (2020)

5. Li, C., et al.: Overview of traffic management of urban air mobility (UAM) with eVTOL aircraft. J. Traffic Transp. Eng. **20**(4), 35–54 (2020)
6. Wei, Q.S., Nilsson, G., Coogan, S.: Safety verification for urban air mobility scheduling. In: IFAC PAPERSONLINE. 2022: 9th IFAC Conference on Networked Systems (NECSYS), pp. 306–311 (2022)
7. Bennaceur, M., Delmas, R., Hamadi, Y.: Passenger-centric urban air mobility: fairness trade-offs and operational efficiency. Transp. Res. Part C Emerg. Technol. **136** (2022)
8. Bulanowski, K., et al.: AURORA-creating space for urban air mobility in our cities. In: Nathanail, E.G., Gavanas, N., Adamos, G. (eds.) CSUM 2022, pp. 1568–1585. Springer, Cham (2023). https://doi.org/10.1007/978-3-031-23721-8_122
9. Koumoutsidi, A., Pagoni, I., Polydoropoulou, A.: A new mobility era: stakeholders' insights regarding urban air mobility. Sustainability **14**(5) (2022)
10. Yavas, V., Tez, O.Y.: Consumer intention over upcoming utopia: urban air mobility. J. Air Transport Manag. **107** (2023)
11. Maia, F.D., Da Saude, J.: The state of the art and operational scenarios for urban air mobility with unmanned aircraft. Aeronaut. J. **125**(1288), 1034–1063 (2021)
12. Perperidou, D.G., Kirgiafinis, D.: Urban air mobility (UAM) integration to urban planning. In: Nathanail, E.G., Gavanas, N., Adamos, G. (eds.) CSUM 2022, pp. 1676–1686 (2023). https://doi.org/10.1007/978-3-031-23721-8_130
13. Lerro, A.: Survey of certifiable air data systems for urban air mobility. In: 2020 AIAA/IEEE 39th Digital Avionics Systems Conference (DASC) Proceedings, 39th AIAA/IEEE Digital Avionics Systems Conference (DASC) (2020)
14. Jha, A., et al.: Urban air mobility: a preliminary case study for Chicago and Atlanta. In: 2022 IEEE/AIAA Transportation Electrification Conference and Electric Aircraft Technologies Symposium (ITEC+EATS 2022). 2022: IEEE/AIAA Transportation Electrification Conference/Electric Aircraft Technologies Symposium (ITEC+EATS), pp. 300–306 (2022)
15. Postorino, M.N., Sarne, G.: Reinventing mobility paradigms: flying car scenarios and challenges for urban mobility. Sustainability **12**(9) (2020)
16. Liberacki, A., et al.: The environmental life cycle costs (ELCC) of urban air mobility (UAM) as an input for sustainable urban mobility. J. Clean. Prod. **389** (2023)
17. Johnson, W.C.: UAM coordination and assessment team (UCAT). In: National Aeronautics and Space Administration (NASA): Ames (2019)
18. Vempati, L., et al.: Assessing human-automation role challenges for urban air mobility (UAM) operations. In: 2021 IEEE/AIAA 40th Digital Avionics Systems Conference (DASC). IEEE/AIAA 40th Digital Avionics Systems Conference (DASC) (2021)
19. Pongsakornsathien, N., et al.: Human-machine interactions in very-low-level UAS operations and traffic management. In: 2020 AIAA/IEEE 39th Digital Avionics Systems Conference (DASC), San Antonio, TX, USA (2020)
20. Straubinger, A., Schaumeier, J., Plötner, K.O.: A roadmap for urban air services. In: Delft International Conference on Urban Air Mobility (2022)
21. Tepylo, N., Straubinger, A., Laliberte, J.: Public perception of advanced aviation technologies: a review and roadmap to acceptance. Prog. Aerosp. Sci. **138** (2023)
22. Johnson, R.A., Miller, E.E., Conrad, S.: Technology adoption and acceptance of urban air mobility systems: identifying public perceptions and integration factors. Int. J. Aerosp. Psychol. **32**(4), 240–253 (2022)
23. Timeline of Drone Integration. Federal Aviation Administration (2022)
24. Chandarana, M., et al.: Streamlining tactical operator handoffs during multi-vehicle applications. In: IFAC Papersonline. 15th IFAC/IFIP/IFORS/IEA Symposium on Analysis, Design and Evaluation of Human-Machine Systems (HMS), pp. 79–84 (2022)

25. Shneiderman, B.: Information-Processing and Human-Machine Interaction - An Approach to Cognitive Engineering - Rasmussen, J.: Information Processing & Management (1), 103 (1988)
26. Work domain analysis: concepts, guidelines and cases. Ergonomics (11), 1790–1791 (2013)
27. Ho, D., Burns, C.M.: Ecological interface design in aviation domains: work domain analysis of automated collision detection and avoidance. In: Proceedings of the Human Factors and Ergonomics Society Annual Meeting, no. 1, pp. 119–123 (2003)
28. Zhang, Y., et al.: Conceptualization of Human Factors in Automated Driving by Work Domain Analysis. SAE Technical Papers (2020)
29. King, B.J.: Identifying risk controls for future advanced brain-computer interfaces: a prospective risk assessment approach using work domain analysis. Appl. Ergon. 104028 (2023)
30. Aminoff, H., et al.: Modeling the implementation context of a telemedicine service: work domain analysis in a surgical setting. JMIR Formative Res. (6), e26505 (2021)
31. Tomitsch, M., et al.: Design. Think. Make. Break. Repeat. A handbook of methods (2018)

User Experience and Inclusivity in MobiTAS

Post-accident Recovery Treatment Using a Physio Room and Aided by an Assistive System

Bertrand David[1(✉)], Martin Reynaud-David[2], René Chalon[1], and Julien Jouffroy[2]

[1] Université de Lyon, CNRS, Ecole Centrale de Lyon, LIRIS, UMR5205, 69134 Lyon, France
{Bertrand.David,Rene.Chalon}@ec-lyon.fr
[2] ANTS, 46-14 Allée d'Italie, 69007 Lyon, France
{martin.reynaud-david,julien.jouffroy}@ants-asso.com

Abstract. Assistive systems can play an important role in a large scope of activities. In this paper, we examine how to assist an injured person in their physical activities in a physio room to allow them to recover their previous physical condition, at least, increase, or maintain their present condition. The objectives of this system are: facilitate for users appropriate training processes and receive information on progress obtained, facilitate for the APA (Adapted Physical Activity) teachers user management, integrate new training devices, increase overall use of the physio room; and, in general, create friendly living conditions for all participants.

Keywords: Digitalization · virtualization · assistive systems · Patient · injuries · surgical treatment · post-operation period · recovery process sequence · User Interface · User Experience

1 Introduction

After a car, bike or scooter injury, the initial period is devoted to medical treatment (and, when necessary, surgical intervention) followed by a passive period dedicated to regeneration of the injured part(s) of the body. After this period and medical treatment, a new active period can start, making it possible to help recovery by appropriate physical therapy. The latter is prescribed by the surgeon and supervised by the physical therapist. This treatment can be efficiently supported by appropriate instruments, tools or machines facilitating physical exercises. The activity can be practiced either in a private physical therapist's office or in a more specialized and amply equipped physical therapist structure (room).

In this paper we will describe an original approach inaugurated in Lyon in 2018, namely the first sports gym dedicated to people with motor disabilities. Once out of hospital environments, it is very difficult for people affected by motor disabilities to maintain daily physical activity, as a result of the absence of venues equipped with accessible devices. The ANTS association S.P.O.R.T. gym (Stimulating People & Organizing

H. Krömker (Ed.): HCII 2024, LNCS 14733, pp. 127–141, 2024.
https://doi.org/10.1007/978-3-031-60480-5_7

Recreational Therapies), proposes a physio room with a large quantity of specialized muscular reinforcement devices [1].

These muscular reinforcement devices are physical-automated equipment allowing several exercise parameters (duration, amplitude, strength, etc.) to be programmed and their results to be collected in order to store them for later use.

To use this kind of physio room appropriately, to facilitate for users the appropriate training processes, and for APA teachers user support, to increase overall use of the physio room, and, in general, create friendly living conditions for all participants, an assistive system appears to be an appropriate solution.

Its objective is to provide more efficient use of several devices specialized in different treatments (practices) and manage exercise parameters and their results in an integrated assistive system. The latter is able to manipulate collected data in order to provide appropriate information for the patient, physical therapist, and surgeon on the progress of the recovery process, and suggestions as to its evolution. Patient reactions, such as treatment appreciation, their perception of recovery progress, and their suggestions, can also be considered.

We had access to the ANTS association S.P.O.R.T. gym physio room with multiple equipment devices (see below), as well as to several patients who discussed with us, in order to observe their behaviors (use, suggestions; and progress of the recovery process), and we collected useful data, which could constitute an assistive system. These observations issued from an initial observation poll allowed us to validate the objective of our approach and led us to design an assistive system that we present in this paper.

The rest of this paper is organized as follows. In Sect. 2, we outline a short state-of-the-art. In Sect. 3, we present the objectives of the ANTS association and its working conditions. In Sect. 4, we clarify the objective of the assistive system and its main components. In Sect. 5, we conclude this study by presenting the next steps in order to obtain a fully operational assistive system.

2 State-of-the-Art

From the medical state-of-the-art point of view, the approach of the ANTS association is relatively new as its objective is to be positioned between the medical context with hospitals and post-operation centers and the physiotherapists and structures available in the city or town. The ANTS room attempts to merge long-term physiotherapist activities in city open space with the possibility for patients to visit frequently and be supervised by APA teachers. This kind of structure is relatively unusual worldwide. We can mention two recently opened in France (Grenoble [2] and Pallavas-les-flôt), while one is currently in use in Casablanca (Morocco).

The main objective of this room is to maintain regular motor activity as recommended by the WHO (World Health Organization) in the CIF (International Classification of Functioning, Disability and Health) [3]. Stimulation of injured limbs through physical activity makes it possible to efficiently fight the risk of secondary complications likely to appear following a spinal injury [4].

Stimulating muscles below the injury allows the cardio-vascular risk to be managed by modifying the ratio between lean body mass/fat body mass, and thus premature risks of carbohydrate and lipidic metabolism dysfunctions [5]. This also reduces thromboembolic risks by repelling venous blood thanks to muscular activity [6], as well as the risk of fracture of the affected limbs. Physical exercise will stimulate loco-regional vascularization, activating osteocyte production and improving dysregulation of bone remodeling [7].

Therefore, it is a means of maintaining both physical and psychological health for the spinal cord injured, through improvement of patient lifestyle and renewed self-esteem. Indeed, the patient will accept a better body image thanks to the maintenance of muscular volume and contour that enhance physical appearance.

These physical transformations also ease daily movements, similar to transfers. Physical activity also ensures for the injured spinal cord a reduction in pain-related complaints and an improvement in mood and well-being by preventing disorders such as anxiety, depression, and isolation. The activities, individual or grouped, and most often recreational, ease the process of social inclusion. In hospitals, accessible gyms are open to patients to allow them to maintain regular physical activity. This practice is essential to compensate the absence of daily physical activity due to the handicap, as it prevents the appearance of secondary injuries listed above (soreness, cardio-vascular disorders, etc.), as a result of paralysis.

3 A Room of Adapted Physical Activities (APA)

The S.P.O.R.T. gym, which means "Stimulating People and Organizing Recreational Therapies", welcomes adults with motor disabilities for the practice of physical activity adapted to everyone. While it is not specifically devoted to post-accident treatment, it may concern people mainly with spinal cord injuries, cerebral injuries or neurodegenerative disease.

The ANTS association adopted a mission to create the missing connections between sports and handicaps. As a primary objective, the association aims to make accessible physical exercise to people affected by a spinal cord handicap.

While for post-accidents the objective is to reach as quickly as possible a pre-accident level, for other categories three situations are to be considered:

1. Increase the initial level and maybe discover new hidden capacities,
2. Maintain as long as possible the current level acting on sporting aspects such as performance. It would seem important for patients to rediscover the pleasure of practice, to surpass them, to meet new people, and to take care of themselves.
3. Try to minimize long-term deteriorations of physical condition by sporting and extra-sporting actions (as mentioned above).

ANTS provides:

- Individual personalized support by APA teachers
- Possibility of more frequent use of this room in comparison with physical therapists' offices
- Association financial support of a grant/allowance to users needing it

Usual working conditions are as follows: Every person registered can practice between 1 and 5 times a week, booking the session beforehand on the website. The sessions are individual, and the APA teacher creates adapted programs according to the goals, capacities, and needs of the person. There are 7 people at most for a one hour session.

The APA team is attentive to each person and ensures an individual follow-up to help each one of them achieve their goals.

The management structure of this organization is the association, and all users of this room are members of it.

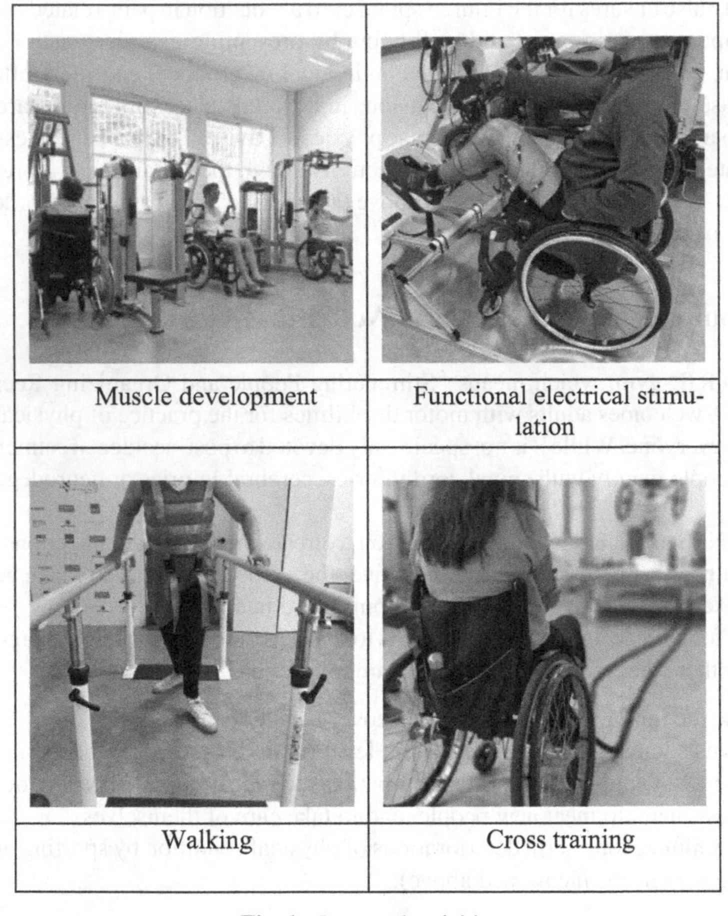

Fig. 1. Proposed activities

Proposed activities are (Fig. 1):

- Muscle development: This activity can have different goals according to the person: strength build-up, maintenance, mobility, autonomy, etc. Many tools are available: bodybuilding machines, handbikes, rowing, small equipment, etc.

- Functional electrical stimulation: Stimulation of the paralyzed limbs with physical activity, creating mobility in the lower part, blood activation, and muscle strengthening. Many tools are available: MOTOmed [8], Bioness orthesis, Bluetens stimulator, etc.
- Walking: Balance and coordination exercises to improve locomotion and mobility of the lower limbs. Many tools are available: parallel bars, walking harness, step, cone, etc.
- Cross training: Muscle development circuit with adapted rhythm and exercises according to the person's capacity. Many tools are available: rack, rope, wallball, medicine ball, weights, etc.

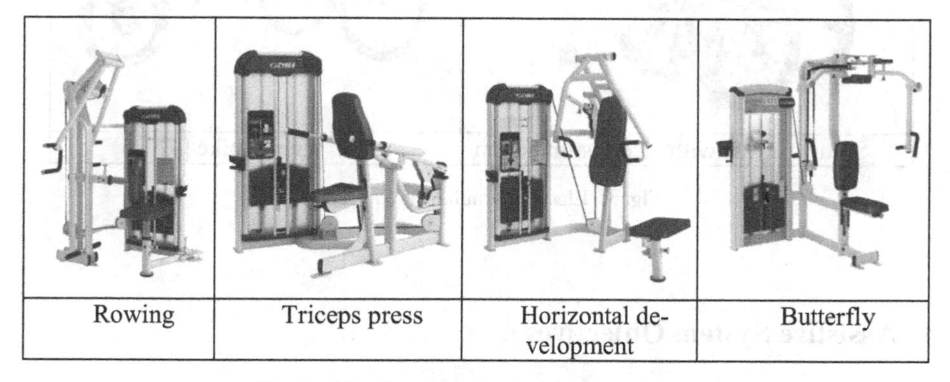

Fig. 2. Muscle strengthening devices adapted

Adapted muscle strengthening devices proposed are (Fig. 2):

- Rowing: Involves back muscles, and participation of rhomboids through arms backwards extension
- Triceps press: Downwards arm extension for triceps contraction and shoulder flexor
- Horizontal development: Involves chest muscles through frontal repulsive force movement
- Butterfly: Involves rhomboids through arms repulsion force, with participation of shoulder blade fixers and posterior deltoids

Electro-stimulation equipment proposed are (Fig. 3):

- Static bicycle with electro-stimulation: This electrostimulation equipped bicycle allows contraction on paralyzed muscles through electric stimulation. The pedaling movement is re-created through non-invasive surface electrodes. An electric current stimulates the motor nerves of the paralyzed muscles in order to provoke muscular contractions, in an order that allows the user to pedal again. The mobilization of the lower limbs through the regular practice of electro-stimulated bicycles improves users' health.
- Berkelbike [9]: The Berkelbike is an electrostimulation bicycle that allows simultaneous pedaling with legs and arms. This bicycle can of course be used outdoors,

but we have set it up in our venue with a home-trainer connected to a software that simulates a movement along an actual path with real-life images on a wide screen.

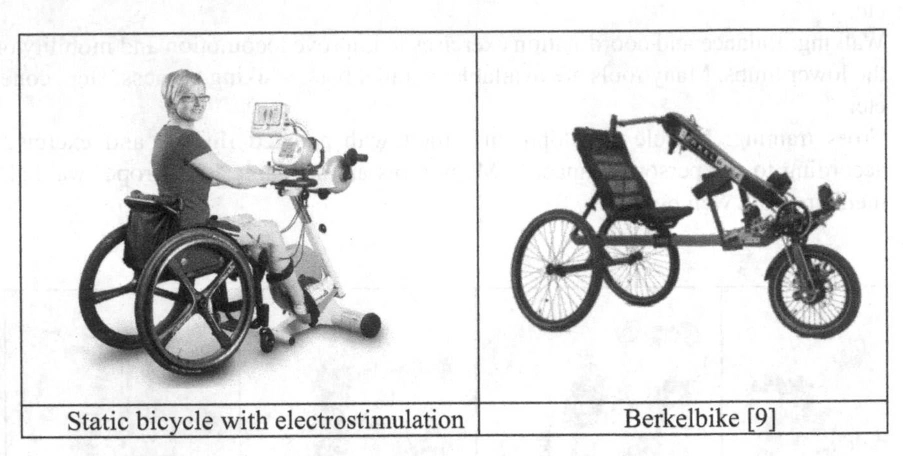

Static bicycle with electrostimulation	Berkelbike [9]

Fig. 3. Electro-stimulation equipment

4 Assistive System Objectives

Availability of this kind of room with the equipment presented above requires the proposal of an appropriate assistive system able to ensure optimum use for all aspects (users' use of equipment, equipment use, etc.).

The objective of this assistive system is to collect human information (professionals and users/patients), to allow physical therapists (APA teachers) to parameterize different equipment devices in relation with prescribed processes for each user, to receive from these devices collected data in order to store them, to produce appropriate presentations facilitating interpretation by human actors in order to allow them to appraise their progress in the recovery process. Algorithmic treatments of collected data could also be proposed in order to consolidate appraisal of the process and its evolution. An algorithmic treatment can also be proposed on these data knowledge bases. To do so, it is important to create a numerical environment able to access and manage all the data needed. This transformation is called digitalization.

4.1 Digitalization Principles

Digitalization is the transformation of current working principles in the physical world to a new world that is more or less digital and virtual [10]. This evolution started at least 15 years ago and has progressed constantly rather than linearly. It can be applied to marketing, to CRM (Customer Relationship Management), to production, to public services, to logistics and so on [11]. While information-based services can be totally digitized, physical object-based activities, production, and manipulation mixing physical and virtual activities with the Internet of Things (IoT) approach [12], are also present. Information digitalization can ensure substantial improvements to data management, thanks to the quality obtained, process time reduction, and associated costs.

Commonly, digitalization is presented as a 3-step process starting with Digitization; transformation of physical documents (texts, drawings, etc.) to digital ones, which can be more easily transported and modified. The next step is known as Digitalization, which is characterized by the presence of manipulation tools (for these digital documents) allowing these data to be shared and presented on different websites. Their transportation between different locations of use, mainly by computer networks and commonly accessible stores, either in specific servers or increasingly more in no-located cloud infrastructures, are thereby facilitated. The last and main step is known as Digital Transformation, which is concerned by the proposals of new organizations considering previously explained elements, leading to what is generally called Informatization [13].

In this way, economic, social, and cultural relationships and barriers are reconsidered and profoundly modified. As stated earlier, methodological processes are not the same as the scope of actions is different. Nevertheless, generic digitalization processes exist and can be used either directly or after adaptation to the field concerned.

Globally, we can define digitalization as a transformation process, the goal of which is to propose: new practices, a new ecosystem, and new management in order to provide better services.

Our paper does not focus on the design process of digitalization in general, as it is devoted to the digitalization of the ANTS Room environment.

4.2 Users' Option Poll

Naturally, we started this project with our perception of this assistive system. However, to increase interest and take into account no-imagined services, we discussed with the ANTS management staff and APA teachers.

To collect the expectations of all users, we used an opinion poll with the following main questions (dominant answer in **bold**, average answer <u>underlined</u>):

- Frequency of your visits to the rehabilitation room (dominant 2).
- Types of exercises or activities performed **(Strength**, **Endurance**, Walking, motor control, Electrostimulation or <u>Circuit training</u> exercises)
- What aspects of your progress would you like to monitor more precisely (**Strength improvement,** <u>Improved mobility</u>, Pain reduction, Improved coordination, Improved distance traveled, <u>improved active time</u> compared to total time, Improved feeling of effort, Reduced fatigue or Other)
- You can specify here what other elements you would like to follow.
- How do you prefer to receive feedback on your progress? **Visually (graphs, tables)**, Verbally (advice, comments), In writing (reports, notes)
- Are there any specific features you would like to see in a tracking tool? Reminders, **Personalized advice**, Integration with medical devices (Doctolib), <u>Personalization of rehabilitation programs</u>, Access to educational resources, Live feedback on form and technique, <u>Logbook: share your condition</u>, your heart rate, etc., Discussion forum with other practitioners, etc.

- Are you comfortable using tracking apps? (dominant Yes)

The results were interesting and for the most part confirmed users' interest in this approach, as summarized previously.

4.3 Assistive System Role and Functions

To provide appropriate services organized in an assistive system, it is important to identify the main aspects of this system. Globally, ITS (Information & Communication Systems) are responsible for considering data collection, storage and management, appropriate manipulations in different working situations, providing communication over short and large distances by wireless and wired networks, in mobility, and interacting with the environment in which active and passive things are organized in the IoT (Internet of Things). Two important aspects are HCI (Human-Computer Interaction) and new services based on AI (Artificial Intelligence) work on collected data. The main HCI objectives are related to multimodal and multi-channel communication with various devices and interaction styles, such as touch-based and speech-based interactions (Chatbots), as well as AR (Augmented Reality) and VR (Virtual Reality) approaches. These interactions can occur either on a desktop computer or with mobile devices (smartphones, connected watches, tablets), allowing individual and collaborative situations, and large, in-the-field, screens or in infrastructure integrated HCI devices. Adapted UI is also to be considered.

4.4 ANTS Assistive System Working Principles

We can schematize the working principles of the ANTS Assistive system as follows (Fig. 4). The injured person or person with disabilities must take out a membership to be able to use the physio room. After a post-operation or other period, medical staff can formulate a diagnosis and prescribe sports activities. This diagnosis is taken into account by room APA teachers, who are able to reformulate it in relation to available devices in the room and draw up a session schema with appropriate exercises and their calibrations (mainly weight and number of repetitions). During the first session, the APA teacher must configure the parameters of each device and register them in the user space for later use. After each exercise, obtained results are collected and stored in the same space. At the end of a session, collected data are consolidated, and an evaluation can be made in order to determine evolution of parameters for the next session. The user can observe and comment this evaluation and formulate personal suggestions expressing their feelings and suggestions for the next session.

We can now briefly present the main elements of the database. Main classes (in object-oriented vision) are:

- User with their characteristics,
- APA teacher with their characteristics and data access rights
- Medical staff
- List of potential exercises with corresponding devices
- List of elaborated sessions based on exercises and used devices

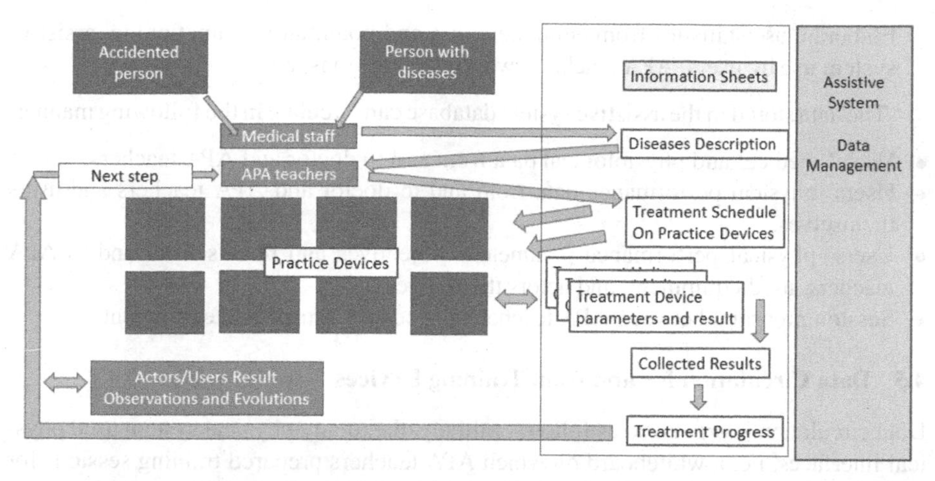

Fig. 4. Organization of the proposed assistive system

- Training device characteristics for initial parametrization for each new user and communication principles for receiving exercise parameters (user specific parameter values, exercise objectives (mainly weight and number of repetitions) and for providing users' results.

Naturally, these classes are characterized by associated attributes and services able to perform calculations, transformations, exchanges with other classes and the User Interface. We can mention the following main attributes for each class and their uses:

- Dispose of appropriate information on users in order to propose activities corresponding to their needs,
- Dispose of all installation, adaptation, use, and recovery information on each training device,
- Be able to access doctors' prescriptions, giving main instructions for the practice of appropriate physical activities,
- Allow APA teachers to elaborate appropriate lists of activities either as potential practices (sessions) or precise descriptions for a scheduled session,
- Collect, store and manage, for each piece of equipment, adaptation parameters taking into account each user's morphological characteristics,
- Collect, store and manage, for each piece of equipment, users' scheduled objectives for a practice session,
- Collect, store and manage, for each piece of equipment' users' results,
- Collect, store and manage users' observations of previous results and their goals for the next practice sessions,
- Take into account new users (association member),
- Take into account new equipment with corresponding parameters,
- Propose appropriate tools allowing the results obtained to be observed: for a session and in a more longitudinal manner,
- Allow APA teachers to annotate users' session results in general or in anticipation of the next session,

- Elaborate use statistics from equipment, use and potential malfunctioning, assistive system use by users, APA teachers' working conditions, etc.

The data stored in the assistive system database can circulate in the following manner:

- Users' medical and physiological data from and to doctor and APA teachers,
- Users' physical performance data from and to doctor and APA teachers and users themselves,
- Users' physical performance parameters, objectives and results from and to APA teachers, used equipment, and actors themselves,
- Session program, between APA teachers, concerned actors, and equipment.

4.5 Data Circulation for and from Training Devices

Data circulation is organized as follows. Initially, the totally physical system used physical interfaces, i.e. a whiteboard on which APA teachers prepared training sessions for users enrolled for the day or half-day. It seems interesting to mention that the first step in digitalization was to provide digital session reservation by users. By digitalized exchange between previously presented classes and their services, it is possible to send or receive requested data either directly or via appropriate API (Application Programming Interface) for remote devices. In a digitalized assistive context, this schedule of activities is visualized on a large display screen located in an appropriate overall visible location. The schedule can be duplicated on APA teachers' tablets, either in full or selectively for a specific user with whom the APA teacher is working. If the device is connected, initial parameters for the current user are transmitted to the device, as well as the working parameters for the exercise, after which the results obtained are sent back to the database. If the connection with the API is not available for the device, the APA teacher proceeds to initial installation and gives the user appropriate instructions on the exercise to be carried out. At the end of the exercise, the APA teacher collects the user's results and transmits them to the system via their tablet. In a totally digitalized version of the assistive system, multiple digital devices (large screen, laptop, mobile computer, tablet or smartphone) will be used, as well as direct connection with training devices providing an appropriate User Interface (Fig. 5).

4.6 Assistive System Platform

To organize all these aspects, it is appropriate to create or use a supportive platform, the aim of which is to connect people, services, and even things in ways that have been unthinkable until now.

In order to organize and manage these aspects, we need a system that groups the different services and their applications, the data on which they work, and the relationship, if needed, between the field and the system. Its architecture is organized in 4 layers (Fig. 6).

These layers (from top to bottom) are the following:

- Layer 1 is devoted to applications,
- Layer 2 is geared towards management of common data and proposing general services.

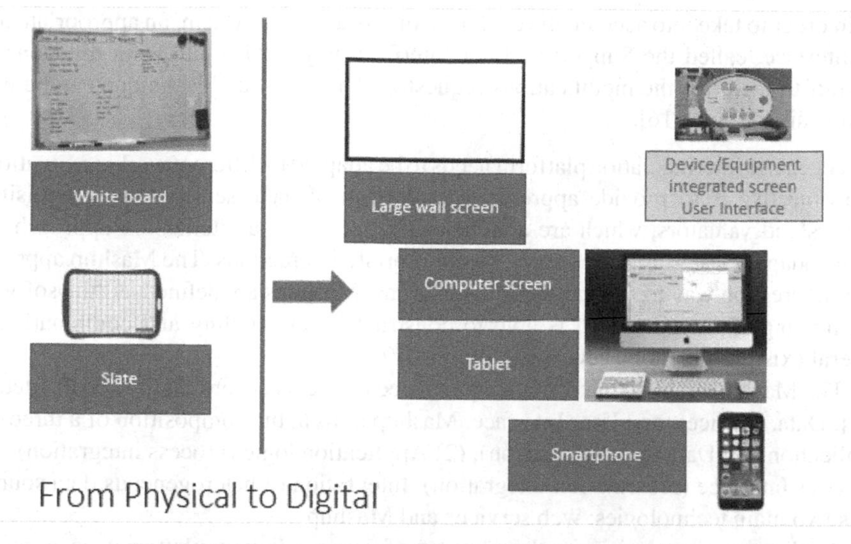

Fig. 5. Evolution of User Interfaces: from physical to digital

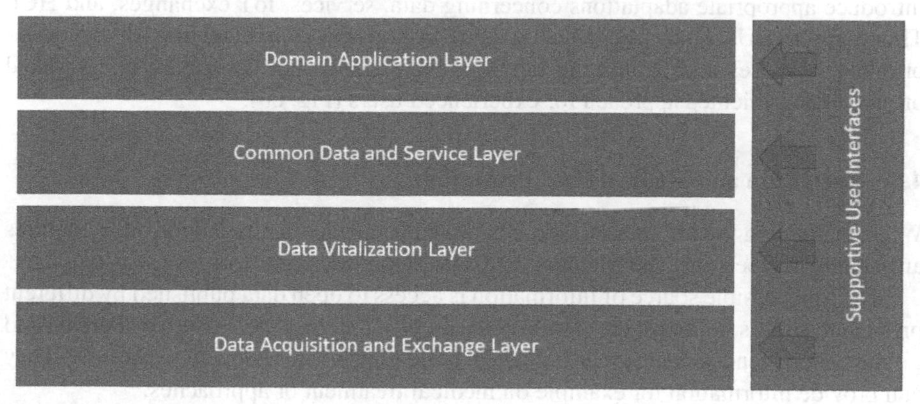

Fig. 6. a/ Architecture of a supportive platform with Data Vitalization [14], b/ Supportive User Interface [15].

- Layer 3 works on what is called *Data Vitalization*. The idea is to work on data and elaborate new constructions on them that can be used in new applications. An example of this need is to be able to collect and associate data in order to manage situations that were not predictable, such as several cases of rescue for not necessarily easily predictable situations.
- Layer 4 aims at acquiring and distributing the data needed for appropriate functioning of all applications concerning the ANTS room.

- In order to take into account the evolution of this assistive system, an appropriate user interface, called the Supportive User Interface, is provided, allowing maintenance staff to facilitate the modifications requested. The "mashup" technique can be used for this purpose [16].

A generic intermediation platform needs to be adapted to different working situations. The objective is to provide appropriate collection of data, services, and acquisition sensors and valuators, which are able to work together in a vitalization approach and create adapted User Interfaces allowing appropriate interactions. The Mashup approach is an interesting way to support these adaptations. Mashups are defined as "the software engineering approach, which is able to construct by assembling and combination of several existing functions new applications" [17].

The Mashup architecture is made up of three elements according to Merrill cited by [17]: Data, Services, and User Interface. Mashup aims at the composition of a three-tier application: (1) Data (data integration), (2) Application logic (process integration), and (3) User Interface (presentation integration). Integration of heterogeneous data sources uses two main technologies: web services and Mashup.

Application mashuping in the context of intermediation platforms is a much-appreciated approach for creating appropriate applications based on reuse of existing ones. It is based on a four-layer architecture (Fig. 6a). To update or adapt data and services, a particular User Interface is used called "Supportive UI", the aim of which is to introduce appropriate adaptations concerning data, services, IoT exchanges, and HCI. This supportive User Interface, which is also Mashup application creation- or evolution-oriented, proposes a programming interface for experienced developers, and a visual programming-oriented approach for experienced users (Fig. 6b).

4.7 Open Data and Medical Data Protection

When examining the information sources, we can consider that two opposite categories are obtained: Open data and Medical data and, in-between, more or less sensitive data.

The first possible source of information is access to open data published by different providers, editors, journals, etc. These data can be either historical data synthetized and giving general tendencies, or real-time data, indicating what is happening recently. They can provide information for example on medical treatment or approaches.

Open Data [18] is the idea that some data should be freely available to everyone to use and republish as they wish, without restrictions from copyright, patents or other mechanisms of control [19]. The goals of the open-data movement are similar to those of other "open(-source)" movements such as open-source software, hardware, open content, open specifications, open education, open educational resources, open government, open knowledge, open access, open science, and the open web. Paradoxically, the growth of the open data movement is paralleled by a rise in intellectual property rights [20].

The second category concerns highly sensitive medical data for which protection is mandatory concerning storage and circulation (respect of private life and medical rules, etc.) In our context, this problem must be studied thoroughly to obtain an appropriate solution based on national, international, and worldwide legislations. At least, we can mention GDPR (General Data Protection Regulation) [21].

4.8 Machine Learning and Deep Learning Use for Collected ANTS Data Study

Prospectively, the ANTS project could benefit from collected data in order to discover hidden behaviors, which can suggest new interpretations and treatment decisions. ANTS had at its disposal data collected manually. With our digitalization approach, we will collect progressively more information that can be studied with Machine and Deep Learning approaches.

Deep Learning is very popular today. The past few decades have witnessed its huge success in many applications. Academia and industry alike have competed to apply deep learning to a wider range of applications due to its capability to solve many complex tasks while providing state-of-the-art results [22]. In some situations, Deep Learning can be used to analyze existing data, leading to interesting solutions. In our context, this approach can be used if we have at our disposal a significant amount of data, which could allow us to work on them in order to discover interesting behaviors.

In practice, we can use the RNN structure or CNN model. For some scenarios with recommendation tasks, we can use multi-layer perceptron, LSTM, and other models [23].

4.9 User Experience (UX)

The term UX (User eXperience) refers to the quality of user experience in any interaction situation. UX describes the overall experience felt by the user when using an interface, a digital device or, more broadly, when interacting with any device or service. UX is therefore to be differentiated from ergonomics and usability.

It was Donald Norman, in the 90s [24], who first used this term "user experience" that would then go on to enjoy the success we know. Donald Norman recalls in a very short and instructive video [25], the origin of the term "UX" and the vision he has of it today, an open vision that makes us challenge certain ideas. Donald Norman and Jacob Nielsen in their NNGroup [26] continue to promote and evangelize this concept with a series of conferences and videos. User Experience can be seen as design and test methods conducted by professional designers, but also as final user tests in working conditions of the application or system [27].

In a digitalization process, such as this one, we can use User Experience to validate proposed user interfaces. As indicated earlier, our users may have particular disabilities that can impact the proposed User Interface. During the UX study, we can detect these problems that can lead us to propose a new design of User Interfaces for this category of users.

5 Conclusion

In this paper, we presented our approach for elaborating an assistive system for a physio room, the objective of which is to increase its usability by injured and long-term disabled users, allowing them to practice as frequently as they wish physio activities supervised by APA teachers. In this way they can either recover their previous physical condition or, at least, increase or maintain their present condition. An opinion poll, expressing the interest of a personal accompanying approach, led us to propose a digitalized environment playing the role of an assistive system. A first prototype (PoC) was developed demonstrating its interest. We can now formulate future orientations for this system, starting with integration of a larger number of devices, offering new kinds of exercises. Integration of new sensors, allowing consideration of not naturally connected devices and individual human body-based data, will also increase the interest of this room.

An open-ended approach for integration of new devices, supporting new activities and associated data, will be developed in order to facilitate this approach by a supportive user interface. Appropriate data visualization techniques will continue to be integrated in order to facilitate appraisal of activity progress on users' physical condition. This accumulation of data naturally leading to integration of data analysis tools based on Big data and AI technics (Machine Learning and Deep Learning) [22, 28]. As mentioned earlier, to upgrade the capacity of longitudinal observation can increase the capacity of these assistive systems. In this context, it also seems appropriate to examine the proposed User Interfaces and observe their adequacy to potential users' disabilities. ANTS is also open to integrate new devices allowing exercises for new disabilities. APA teachers are able to take into account cognitive problems.

Last but not least, evolution is the integration of real-time control of the devices, allowing dynamic evolution of their behaviors, such as road profiles for bicycles or home trainers [29].

In this way, the objective to increase friendly living conditions for all users/participants of the ANTS room will be achieved.

Acknowledgements. This work is a collaboration between an academic team from Ecole Centrale Lyon and an association ANTS in charge of managing a physio room. We thank the Centrale Digital Lab and the Ecole Centrale de Lyon students for their contributions to PoC elaboration, as well as the ANTS APA teachers and ANTS users for their interest in this project.

The authors are grateful to the ASLAN project (ANR-10-LABX-0081) of the Université de Lyon, for its financial support within the French program "Investments for the Future" operated by the National Research Agency (ANR).

References

1. https://ants-asso.com/en/home/. Accessed 25 Jan 2024
2. https://ants-asso.com/sport-grenoble/en/s-p-o-r-t-en/. Accessed 25 Jan 2024
3. https://www.who.int/. Accessed 25 Jan 2024
4. Crane, D.A., Hoffman, J.M., Reyes, M.R.: Benefits of an exercise wellness program after spinal cord injury (2015). https://pubmed.ncbi.nlm.nih.gov/26108561/
5. Evans, N., Wingo, B., Sasso, E., Hicks, A., Gorgey, A., Harness, E.: Persons with Spinal Cord Injury (2015). https://www.archives-pmr.org/article/S0003-9993(15)00118-5/fulltext
6. Rimaud, D., CBoissier, Ch., Calmels, P.: Evaluation of the Effects of Compression Stockings Using Venous Plethysmography in Persons with Spinal Cord Injury (2008). https://www.tandfonline.com/doi/abs/10.1080/10790268.2008.11760713
7. Chang, X., Xu, S., Zhang, H.: Regulation of bone health through physical exercise: mechanisms and types (2022). https://www.ncbi.nlm.nih.gov/pmc/articles/PMC9768366/
8. https://www.motomed.com/en/?lang=1. Accessed 25 Jan 2024

9. https://berkelbike.com/. Accessed 25 Jan 2024
10. El Saddik, A.: Digital twins, the convergence of multimedia technologies. In: IEEE MultiMedia, April–June 2018, IEEE Computer Society (2018). https://doi.org/10.1109/MMUL.2018.023121167
11. Abdel-Aziz, A.A., Abdel-Salam, H.A., El-Sayad, Z.: The role of ICTs in creating the new social public place of the digital era. Alexandria Eng. J. **55**(1), 487–493 (2016)
12. Balandina, E., Sergey Balandin, S., Koucheryavy, Y., Mouromtsev, D.: IoT use cases in healthcare and tourism. In: Proceedings of 2015 IEEE 17th Conference on Business Informatics (2015)
13. Firican, G.: What is the difference between digitization, digitalization, and digital transformation? https://www.lightsondata.com/what-is-the-difference-between-digitization-digitalization-and-digital-transformation/. Accessed 05 Jan 2023
14. Xiong Z.: Smart city and data vitalisation. In: The 5th Beihang Centrale Workshop, 7 January 2012
15. David, B., Chalon, R.: Orchestration modeling of interactive systems. In: Jacko, J.A. (ed.) HCI 2009. LNCS, vol. 5610, pp. 796–805. Springer, Heidelberg (2009). https://doi.org/10.1007/978-3-642-02574-7_89
16. Grammel, L., Storey, M.: An end user perspective on mashup makers. University of Victoria Technical Report DCS-324-IR (2008)
17. Atrouche, A., Idoughi, D., David, B.: A mashup-based application for the smart city problematic. In: Kurosu, M. (ed.) HCI 2015. LNCS, vol. 9170, pp. 683–694. Springer, Cham (2015). https://doi.org/10.1007/978-3-319-20916-6_63
18. Auer, S., Bizer, C., Kobilarov, G., Lehmann, J., Cyganiak, R., Ives, Z.: DBpedia: a nucleus for a web of open data. In: Aberer, K., et al. (eds.) ASWC/ISWC-2007. LNCS, vol. 4825, pp. 722–735. Springer, Heidelberg (2007). https://doi.org/10.1007/978-3-540-76298-0_52
19. Kitchin, R.: The Data Revolution, p. 49. Sage, London (2014). ISBN 978-1-4462-8748-4
20. https://en.wikipedia.org/wiki/Open_data. Accessed 25 Jan 2024
21. https://gdpr-info.eu/. Accessed 25 Jan 2024
22. Paul, C., Jay, A., Emre, S.: Deep neural networks for youtube recommendations. In: Recsys, pp. 191–198 (2016)
23. Huang, P.-S., He, X., Gao, J., Deng, L., Acero, A., Heck, L.: Learning deep structured semantic models for web search using click through data. In: CIKM 2013: Proceedings of the 22nd ACM International Conference on Information & Knowledge Management, pp. 2333–2338 (2013)
24. Norman, D. in the 90s. http://mickopedia.org/mickify?topic=Experience_design. Accessed 05 Jan 2023
25. Video Norman's UX presentation. https://youtu.be/9BdtGjoIN4E. Accessed 05 Jan 2023
26. NNGroup. https://www.nngroup.com/. Accessed 05 Jan 2023
27. Nielsen, J.: Keynote the Immutable Rules of UX. https://www.youtube.com/watch?v=OtBeg5eyEHU. Accessed 05 Jan 2023
28. Tsai, C.W., Lai, C.F., Chao, H.C., et al.: Big data analytics: a survey. J. Big Data **2**, 21 (2015). https://doi.org/10.1186/s40537-015-0030-3
29. https://eu.zwift.com/fr/products/runn?variant=42781150445819. Accessed 25 Jan 2024

Research on Electric Vehicle Charging Pile Design Based on Kansei Engineering and Textual Sentiment Analysis

Miao Liu and Ronghan Yang(✉)

East China University of Science and Technology, Shanghai 200237, People's Republic of China
15754303911@163.com

Abstract. The article aims to explore a better way of user needs understanding and a rapid prototype design in product iteration and development. Several textual sentiment analysis methods such as HowNet dictionary, BERT and KMeans was integrated into kansei engineering to recognize the kansei words more precisely. The design process proposed by the study was used in electric vehicle charging pile design. As an affiliated product in the car industry, the charging pile design iterates quickly, therefore a faster way to understand customers with less resource consumption is needed. By using the core HowNet dictionary, the study established a kansei word dictionary from online user reviews and articles. The user reviews and articles were also processed by BERT pretrained model and clustered by KMeans to show more information of the text, in case of some key words missing when using HowNet independently. Then, 10 kansei word pairs were defined to represent the user' sentiment towards the existing charging piles. Finally, semantic differential method was used to evaluate the existing piles design to scope the users' preferences and lead the prototype design. Following the preferences of users, three prototypes were designed to validate if the sentiment of users were correctly understood. The study shown a rapid design process that is effective in the development of products like charging piles. The feasibility and accuracy in Chinese kansei word extraction by HowNet dictionary were also validated.

Keywords: Charging Pile design · Textual Sentiment Analysis · HowNet Dictionary · Kansei Engineering

1 Introduction

1.1 Background

The automobile industry is now evolving rapidly, by the use of new energy from batteries, a new era for mobility is coming, followed by different design trend for vehicles and its related products. More than 100 electric vehicle manufacturers in China are competing severely. Behind the industry, the power recharging plays an important role in the competition of overall experience. Well-designed charging pile will help automobile companies in providing high quality and consistent user experience, which is crucial in

H. Krömker (Ed.): HCII 2024, LNCS 14733, pp. 142–155, 2024.
https://doi.org/10.1007/978-3-031-60480-5_8

the use of electric cars. Charging pile design has critical influence on the recognition of a car brand. The car design is highly related to its charging pile. The car brands, especially those who are pursuing a high-end niche in the market, need to improve their competence by highly systemic design. Therefore, the charging pile and vehicle should follow a similar design style or concept.

However, unlike the car, the charging pile developing period is rather short. Due to this, it is important to understand the user needs and expectations as fast as possible. The design process should be finished quickly and the iteration may be more frequent, so, there should be a simple method to rapidly prototype a new design. Conventional design process of such devices usually require more expertise involved, and it is hard to converge the car style and charging pile style. The charging pile is designed separately. Therefore, three goals of charging pile design concluded here: 1) better understanding and transferring of user needs in a short time; 2) less human resource (design expertise) consuming; 3) creating a more consistent branding experience by following the car design.

1.2 Literature

Seldom study focused on the charging pile exterior design. As a kind of affiliated product of the vehicle power system, the charging pile was ignored before. However, kansei engineering and textual sentiment analysis can be applied in the field to help establish a systemic and fast design methodology. Kansei engineering had been used in car and charging pile design for a long time [1]. In the past it was fuel car and now electric. Though kansei engineering is still effective in design, the electric car design is different from the fuel car, the first is to correctly recognize user needs [2]. In traditional fuel car design, opinions from experts are required in the kansei engineering process. However, the opinions are limited on a small scale. And experts are inevitable to be trapped in stereotypes [3]. It is necessary to improve the kansei design process to recognize user needs faster and precisely. By the involving of textual sentiment analysis based on user generated contents on the product community, the kansei words are easier to be extracted and validated.

Millions of words generated by users in the digital world has been important corpus for researchers to scope the user needs. By using of NLP (Natural Language Process) toolkits, researchers can elicitate critical touch points and expectations of users. Ben explored a mass scale reviews on a Chinese e-market—Jingdong—to provide a method for designers to predict the customer needs, in which some simple NLP methods were used, such as word frequency calculation, word cloud visulization, and NLPIR sentiment analysis [4]. Cong proposed a small sample data-driven method by the application of SIFRanking and pretrained language model to process multiple online review texts, which can help in product iteration [5].

Jin proposed a kansei integrated KANO model to fulfill the process of user demand mining. By syntactic calculating and clustering algorithms, affective emotions and related product features were shown [6].

Based on researches reviewed above, the article aims to better investigate user needs for charging piles design and to explore a systemic rapid design process by use of textual sentiment analysis and kansei engineering.

2 Method

2.1 User Review Text Processing

From reviews and articles on social media and online car communities, researchers extracted the perception of the public towards new energy vehicles and charging piles to scope their expectations. These texts were then collected together, processed in python for additional analysis. Since the tests were all Chinese, jieba was used to split the words, and the meaningless words were removed from the texts.

2.2 Textual Sentiment Analysis

For Chinese textual analysis, the HowNet dictionary and bert was used. HowNet is a widely used Chinese dictionary based on sememe developed by THUNLP. Each concept in HowNet is annotated with a sememe-based definition, the POS tag, sentiment orientation, example sentences, etc., which provides an authentic way to deal with Chinese NLP tasks [7].

Researchers here constructed a kansei words dictionary from the core HowNet dictionary downloaded from its website [8]. Information in the dictionary was initially organized in a txt file, then it was formulated into a csv file. The word name, POS tag and sentiment orientation were arranged in columns and rows, so the dictionary was easy to query in python. Then the words in reviews and article texts can be judged and the kansei words can be extracted [9, 10].

As a pretrained language model, BERT was recently used in many NLP tasks especially for the sentiment analysis on social media to investigate the public opinions [11]. Here it was used to extract and classify the kansei words. The BERT pre-trained model can process a great amount of data with little time consumed. It helps designers to further explore kansei words, and avoid missing them.

Words representing the users' affection and expectation were elicitated, sorted by frequency and classified by K-means. The TD-IDF was used to vectorize the words when applying BERT.

2.3 Kansei Engineering Application

In the article, kansei engineering was used to assist design. Kansei Engineering is a classical method in design. It translates users' perceptual needs into quantitative data, forming a mapping relationship with the product form so that the product shape better meets users' needs [12]. For decades, KE was applied in vehicle design, and it was continually integrated with other design models. Semantic differential is one of the classical methods been used in KE. The semantic differential approach transforms users' perceptual perceptions into quantitative data through an equivalence scale [13]. It is a quantitative research approach used to measure individuals' perceptions of products, services, or concepts. This method involves having respondents evaluate the sensory attributes of a target by choosing a position between opposing semantic extremes. It is employed to gain subjective insights into users' preferences and perceptions, aiding in a better understanding of user needs.

In this paper, kansei words was not generated by designer's subjective judgement, however, it was obtained from the textual analysis from real user talking on the internet. By additional analysis of the sentiment words, which are mostly adjectives and adverbs, the kansei words pairs were confirmed. Then, by semantic differential method, it was used to estimate the current products through questionnaire.

The questionnaire was a 7-point likert scale, providing existing products collected online as samples. To a given sample, the subjects were asked to tell their feeling about the degree the sample fit to the kansei words, and their overall satisfaction score. From the user's feedback, the current designs and kansei words were ranked, so the expectations of customers can be perceived, the design direction can be confirmed.

2.4 Validation

To estimate if the kansei words and design direction were correctly defined, researchers designed charging piles following the style of samples with high scores. The words that were highly scored in those samples were used to evaluate the design. A similar questionnaire with a 7-point likert scale was also used here.

3 Experiment

How he experiment be done was illustrated by a flowchart (see Fig. 1).

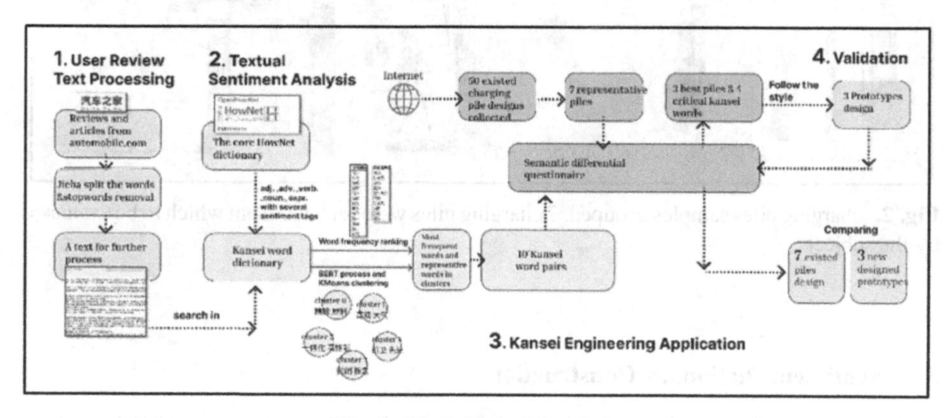

Fig. 1. Flowchart of the study.

3.1 Data Collection

The online review and articles were collected by python request from autohome.com (汽车之家), the biggest vertical portal website about cars in China. The reviews and articles containing charging piles and car styling were collected, aggregated to the corpus in a text. Then, a corpus of 61217 words was established.

3.2 Existing Pile Design Samples Collection

First, 50 samples were collected from electric automobile manufacturers and the design websites. 34 of them were chosen and divided into 7 groups by the author in the next step. Last, several experts in industrial design and visual communication design picked out 7 samples that are representative in the 7 groups, respectively. To minimize the influence of color, the color of these samples' pictures were all removed. The pictures are also cut into similar sizes. Figure 2 shows the grouped samples.

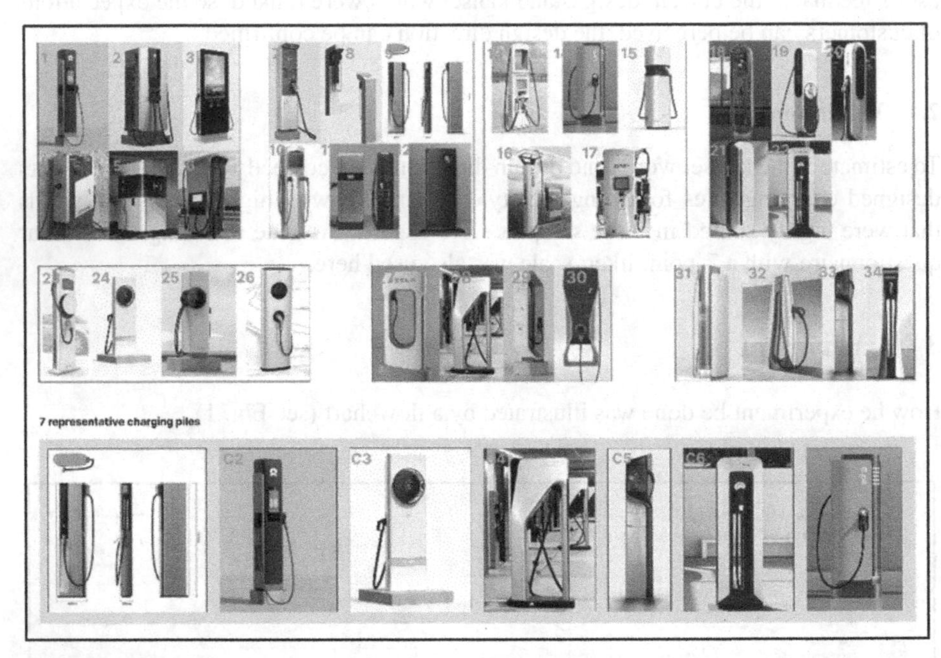

Fig. 2. charging piles samples grouped. 7 charging piles were selected from which to be evaluated by the subjects.

3.3 Sentiment Dictionary Construction

In the core HowNet dictionary, each word contains the information of its POS tag, sentiment orientation. The sentiment orientation contains PlusAttr, PlusEvent, PlusFeeling, PlusSentiment, MinusAttr, MinusEvent, MinusFeeling, MinusSentiment, which means positive attribute, positive event, positive feeling, positive sentiment, negative attribute, negative event, negative feeling, negative sentiment, respectively. According to the demand for kansei word extraction, the adjectives and adverbs with sentiment orientations of PlusAttr, PlusFeeling, PlusSentiment, MinusAttr, MinusFeeling, MinusSentiment were set as constraints on the choice of kansei words. To construct a HowNet dictionary suitable for the kansei word extraction, a text of user review was prepared by jieba words splitting and stopwords removal at first. Second, the words in the text were searched in the core HowNet dictionary. If it was in the core Hownet dictionary and satisfies the constrains, it was chosen as kansei word. The kansei word chosen were added together to form a kansei word dictionary. Table 1 shows the structure of the dictionary.

3.4 Kansei Words Extraction

The text and sentiment dictionary were input into python, Each word in the text was searched in the dictionary and its frequency was counted when it was find in the dictionary. The word was added to positive words list if the sentiment is positive. In contrast, it was added to the negative words list. The words with the first 30 highest frequency in each list were chosen to show in Table 2. For deeper understanding of the sentiment of users, more positive and negative words including verbs, nouns, exprs were picked put to establish a corpus, it was further processed by bert and classified by K-Means. The corpus was classified into 5 clusters. The silhouette coefficient of K-means is 0.79996, which signifies a good categorization [14]. Table 3 shows the representative words in five clusters obtained from BERT and KMeans. Then, 10 positive sentiment words were carefully selected by referring to Table 2 and the opinions from experts. The most frequent words are selected, and some words are integrated into one word, for example, "sharp, cutting-edge, precise" were explained by "technological". A kansei word pair was formed by a positive sentiment word and a negative sentiment word, which is selected from the negative words list or defined by the author according to its semantic. Table 4 is the information of kansei word pairs.

Table 1. The structure of sentiment dictionary.

Chinese Words	English words	POS tag	Sentiment orientation
简洁	Laconic	Adj	PlusSentiment|正面评价
脆弱	Fragile	Adj	MinusSentiment|负面评价
低调	Low-keyed	Adj	PlusAttr|正面属性
病态地	morbidly	Adv	MinusAttr|负面属性

3.5 Kansei Evaluation of the Samples

In the article, Semantic Differential Method was used to evaluate the existing charging pile designs. Users' affection and satisfaction towards specific design were evaluated. Researchers designed a questionnaire to see the user's feelings towards 7 charging piles. The questionnaire examined the opinions about how a kansei word fits a charging pile and the satisfaction score of each pile. In this way, it was learned which kansei words are most important for user satisfaction. Thus, the design direction was also confirmed. Figure 3 shows the structure of the questionnaire. The questionnaire also contains the basic demographic information, such as sex, age, and education (see Fig. 4). The respondents were also asked if they have known or used the charging piles before, to make sure they have basic understanding of the product.

Table 2. The first 30 sentiment words found from sentiment dictionary.

positive words			negative words		
Chinese Words	English Words	Frequency	Chinese Words	English Words	Frequency
系统	systemic	55	封闭式	Enclosed	18
豪华	luxurious	27	混合	Hybrid	5
前卫	Trendy	19	概念化	Conceptualized	4
简洁	Simple and Concise	17	被动	Passive	3
流畅	smooth	16	奢华	Luxurious	3
高端	High-end	16	陌生	Unfamiliar	3
一体化	Integrated	15	虚拟	Virtual	2
强大的	Formidable	13	凹凸	Concave and convex	2
简约	brief	13	特立独行	Eccentric	2
高性能	High-performance	12	挑剔	Picky	2
硬朗	Tough	12	木	Wood	2
依旧	Still	11	堪忧	Alarming	2
宽大	Spacious	10	性感	Sexy	2
精致	Exquisite	9	凶狠	Fierce	1
犀利	Sharp	9	错综复杂	Intricate	1
高级	Advanced	7	迷糊	Confused	1
先进	Cutting-edge	7	吃力	Laborious	1
特别	Special	7	漂浮	Floating	1
大气	Grand	6	繁杂	Complicated	1
超级	Super	6	细碎	Fragmented	1
大胆	Bold	6	步履维艰	Arduous	1
优雅	Elegant	6	斜	Slanting	1
高效	Efficient	5	灰色	Gray	1
透明	Transparent	5	隐约	Indistinct	1
紧凑	Compact	5	若隐若现	Vague	1
抢眼	Eye-catching	5	脆弱	Fragile	1
凌厉	Sharp	5	超然	Transcendent	1
精准	Precise	5	崎岖	Rugged	1
漂亮	Beautiful	4	牛	ferocious	1
舒适	Comfortable	4	异形	Alien	1

Table 3. The sentiment words processed by BERT and classified by KMeans into 5 clusters.

Clusters	Sentiment Words
Cluster 0	系统, 豪华, 简洁, 流畅, 简约, 硬朗, 依旧, 精致, 犀利, 身份, 先进, 特别, 超级, 优雅, 乐趣, 紧凑, 优势, 亮点, 抢眼, 凌厉 Systemic, Luxurious, Simple and Concise, Smooth, Pure, Hard, Still, Refined, sharp, Identity, Advanced, Special, Super, Elegant, Fun, Compact, Edge, Highlights, Eye-catching, Stern
Cluster 1	高端, 强大, 创新, 吉利, 宽大, 高级, 大气, 大胆, 高效, 透明, 混合, 环保, 清晰, 平滑, 成熟, 优秀, 真实, 事实上, 不俗, 轻松 High-end, Powerful, Innovative, Jilly, Generous, Premium, At-mospheric, Bold, Efficient, Transparent, Hybrid, Eco-Friendly, Clear, Smooth, Mature, Excellent, Real, Factual, Uncommon, Relaxing
Cluster 2	动力, 前卫, 方正, 出众, 醒目, 活力, 原创, 倾力, 分明, 信心, 魅力, 清新, 本质, 吃力, 崎岖, 暴力, 费力 Power, Avant-garde, Square, Outstanding, Eye-catching, Energet-ic, Original, Leaning, Distinctive, Confident, Charming, Fresh, Essence, Earnest, Rugged, Violent, Feeble
Cluster 3	封闭式, 概念化, 可圈可点, 必不可少, 尽可能, 显而易见, 风起云涌, 淋漓尽致, 独具特色, 豪华型, 孜孜不倦, 满足感, 颠覆性, 错综复杂, 步履维艰, 若隐若现 Closed, Conceptualized, Distinguishable, Indispensable, As far as possible, Obvious, Windy, Dripping, Distinctive, Luxurious, Tire-less, Satisfying, Subversive, Intricate, Striding, Hidden
Cluster 4	一体化, 高性能, 与众不同, 有意思, 特立独行, 引人注目, 独具匠心, 流畅地, 大功率, 小巧玲珑, 高精度, 少不了, 了不起, 一目了然, 高强度, 游刃有余, 捉摸不定 All-in-one, High-performance, Distinctive, Interesting, Maverick, Striking, Ingenious, Fluent, High-power, Compact, High-precision, Indispensable, Terrific, Obvious, High-strength, Easy to handle, Elusive

Table 4. Kansei word pairs.

	kansei word pairs				
	Chinese	English		Chinese	English
1	科技的	Technological	-	传统的	Traditional
2	系统的	Systemic/Integrated	-	零散的	Scattered
3	高端的	High-end	-	低端的	Low-end
4	豪华的	Luxurious	-	便宜的	plain
5	前卫的	Trendy	-	落后的	Outdated
6	简洁的	Simple and Concise	-	繁复的	Complicated
7	流畅的	Smooth	-	生硬的	Rigid
8	高性能的	High-performance	-	脆弱的	Fragile
9	精致的	Exquisite	-	粗陋的	Rough
10	优雅的	Elegant	-	庸俗的	Vulgar

The data collected from questionnaire was input into the SPSS. The reliability and validity of the questionnaire were examined and, for every pile, its average score of 10 different sentiment words was calculated.

3.6 Prototype Design and Validation

From the analysis of questionnaire, researchers find the best three scored piles and the most influential kansei words for these piles. Then the author extracts design elements from those piles and follows the feeling of the important kansei words to make three new design prototypes (see Fig. 5). Next, these new designs were evaluated by a similar questionnaire in the same population to see if it is better than the existing piles scored before.

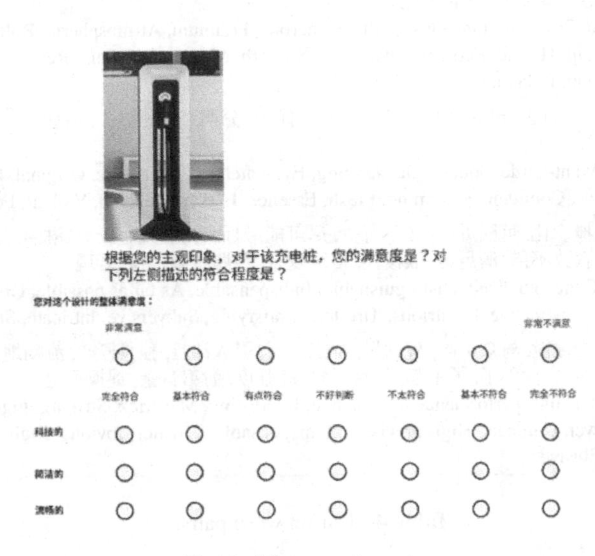

Fig. 3. The questionnaire.

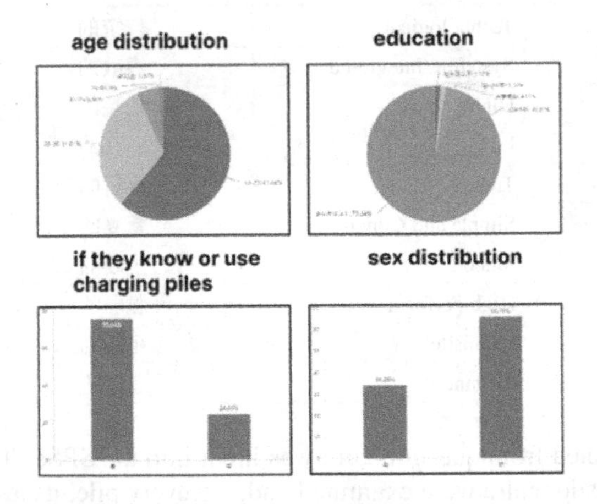

Fig. 4. Demographic information of the respondents.

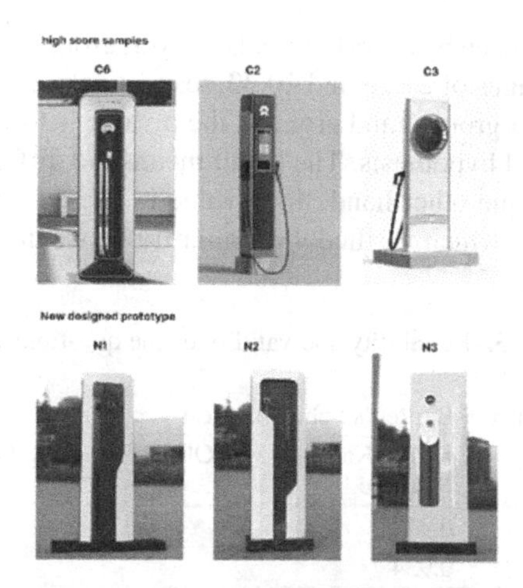

Fig. 5. Following the high score samples style and high score Kansei words, three new designed piles were prototyped by sketch,3D modeling and keyshot rendering.

4 Results

4.1 Satisfaction and Kansei Scores of the Existing Piles

100 questionnaires were distributed, 85 of them were collected, among which 73 were effective. The respondents were both told the basic background information about the experiment. The questionnaires with too short a response time, as well as those with identical answers were excluded. The demographic information shows that 93% of the respondents age between 18 and 30. 75% of them have a master degree. 75% know or use charging pile, and, 65% are female (see Fig. 4).

For every sample, the reliability was examined by the calculation of Cronbach Alpha, the validity was verified by KMO value and Bartlett coefficient. Table 5 shows the outcome. Both the Cronbach Alpha and KMO value are larger than 0.9, that means the data is reliable.

The average and total score was shown in Table 6. The samples were ranked. Sample C6, C2, C3 have significantly higher scores than other samples. For the three samples, technological, systemic and integrated, simple and concise, smooth are both the first four ranked kansei word pairs. So the design will follow these samples and kansei words.

4.2 Design Validation

50 questionnaire were collected for the validation, in which the respondents belong to the same population. Table 7 shows the scores of different samples for 4 chosen kansei word pairs. In the table, group 1 are the three best samples with higher scores; group 2 is the prototype designed by the author.; group 3 is the four worst samples with lower scores. Group 1 and group 2 have similar mean scores, total averages are 21.38 and 21.28 respectively, that means the prototypes designed by the author are as popular as

the best three samples chosen by users before. However, group 2 is greatly different from group 3, with total averages of 21.28 and 20.33, respectively. It can be verified by t-test in Table 8. In the t-test of group 2 and group 3, the p value is lower than the alpha value = 0.05, so refuse the null hypothesis. The result means the difference between group 2 and 3 is significant. On the other hand, the p value in t-test of group 1 and group 2 is larger than alpha = 0.05, which verified the similarity between these two groups.

Table 5. Reliability and validity of the questionnaire.

Samples	Cronbach's Alpha	Bartlett's Sphericity Test	
		KMO (Kaiser-Meyer-Olkin) Sampling Adequacy Measure	Significance
C1	0.934	0.912	0.000
C2	0.954	0.934	0.000
C3	0.945	0.901	0.000
C4	0.954	0.901	0.000
C5	0.95	0.901	0.000
C6	0.96	0.889	0.000
C7	0.951	0.887	0.000

Table 6. Average Scores of different samples for 10 kansei words pairs.

Kansei words pairs	Samples						
	C6	C2	C3	C5	C1	C7	C4
科技的 **technological**	**5.74**	**5.40**	**5.52**	5.12	5.07	5.11	5.37
系统的 **Systemic and integrated**	**5.73**	**5.63**	**5.45**	5.30	5.34	5.51	4.93
高端的 high-end	5.68	5.30	5.29	5.05	5.15	4.93	4.93
豪华的 luxurious	5.33	4.95	5.08	4.93	4.53	4.92	4.73
前卫的 trendy	5.37	4.96	5.29	4.86	4.85	4.63	5.21
简洁的 **simple and concise**	**5.74**	**5.55**	**5.63**	5.16	5.41	5.23	5.07
流畅的 **smooth**	**5.66**	**5.59**	**5.47**	5.21	4.97	4.97	5.07
高性能的 high performance	5.56	5.41	4.99	5.00	5.25	5.14	4.99
精致的 exquisite	5.45	5.32	5.25	4.96	4.78	4.92	4.74
优雅 elegant	5.42	5.40	5.26	4.89	4.93	4.78	4.85
total score	**55.68**	**53.49**	**53.22**	50.49	50.29	50.14	49.88

Table 7. Scores of different samples for 4 kansei words pairs.

group	Samples	简洁的	系统的	流畅的	科技的	Total	Total Average
1	C6	5.85	5.58	5.73	5.46	22.62	21.38
	C2	5.04	5.58	4.96	4.92	20.50	
	C3	5.50	4.85	5.46	5.23	21.04	
2	N1	5.81	5.19	5.31	5.38	21.69	21.28
	N2	5.12	5.31	4.96	5.27	20.65	
	N3	5.58	5.35	5.15	5.42	21.50	
3	C1	5.07	5.24	5.12	4.97	20.40	20.33
	C4	5.27	4.97	5.17	4.95	20.36	
	C5	5.12	5.01	4.86	5.06	20.05	
	C7	5.11	5.21	5.23	4.97	20.52	

Table 8. t-test for the mean of different group samples.

	Group 2	Group 3	Group 1	Group 2
Mean	5.320513	5.083707	5.346154	5.320513
Variance	0.048772	0.014534	0.112695	0.048772
Observation Values	12	16	12	12
Pooled Variance	0.029019		0.080733	
Hypothesized Mean Difference	0		0	
df	26		22	
t Stat	3.640162		0.221047	
P(T <= t) one-tailed	0.000593		0.413547	
t Critical value for one-tailed test	1.705618		1.717144	
P(T <= t) two-tailed	0.001186		0.827095	
t Critical value for two-tailed test	2.055529		2.073873	

5 Discussion

5.1 The Use of HowNet Dictionary

The researchers used the HowNet dictionary in the study to search for kansei words. In this way, the kansei word extraction should be more objective than traditional process. However, the effect of the application might be limited by the data collection. The amount of text and the website have a great influence on the reliability of the data. For deeper minging of the big data, the HowNet dictionary should be used in other ways, for example, the semantic similarity computing will help the researchers to understand the structure of kansei word, thereby the words can be better concluded [15, 16]. Since the HowNet provide a method for semantic similarity calculation, the sentiment orientation can be shown in a quantitative form. Thus, it might be better to validate the design by the sentiment value calculated from HowNet.

It is possible that the BERT and KMeans can be used to categorize the kansei words, instead of using the PCA to conclude. There is another challenge for the NLP process since the NLP depends on mass amount of data, that means the products without consumer reviews are hard to analyze. In this study, the car design served as a heterogenous product that is highly decisive for charging pile design, so the reviews about car design are also incorporated into the corpus.

In kansei word extraction, the word frequency was the most important factor being considered. There might be more structural method to do the same job by combing the sentiment value computing and text categorization. And, the visualization of user affection might be effective in the design iteration.

5.2 The Design Process Optimization

The three piles designed by the author are significantly better than other piles designed before. Though the design process is quite short, it has confirmed that the kansei words and samples are correctly selected. However, kansei engineering is a systemic design methodology. More works should be done in the prototype design and its validation. Other methods such as GRA [17], Grey-Markov model [18] can also be applied to supplement the process in kansei engineering to recognize users' requirements and product forms.

6 Conclusion

The study tried to explore a rapid and systemic methodology for product development. By applying HowNet dictionary and some other NLP methods, it has enhanced the kansei word recognition in a short time, which is helpful for the user needs definition, the product evaluation and iteration. Well organized kansei words and reliable questionnaires provided clear understanding about the product. For some products like charging piles that iterate rapidly in the market, the method proposed by the study can balance well between rapid development and objective understanding of user needs.

The study was done on a small scale of samples. However, it is effective. It demonstrates that in digital world, user generated contents can reduce the need for offline samples. For products require long period of design, it might be more reliable if the experiment be done on a larger social media for a longer time.

The textual sentiment analysis was powerful in user affection analysis since websites and social network are manufacturing mass amount of information. By incorporating kansei engineering and textual sentiment analysis, the study shows a probability for deeper integration of digital computing and industrial design. And, less human resources and time consumed in design.

Disclosure of Interests. The authors have no competing interests to declare that are relevant to the content of this article.

References

1. Ding, Y.: Research on form design of electric bicycle charging station according to kansei engineering. Appl. Mech. Mater. **437**, 296–300 (2013)
2. Zhang, F., Wang, J.: Application of kansei engineering in electric car design. Appl. Mech. Mater. **437**, 985–989 (2013)
3. Lai, X., Zhang, S., Mao, M., Liu, J., Chen, Q.: Kansei engineering for new energy vehicle exterior design: an internet big data mining approach. Comput. Ind. Eng. **165**, 107913 (2022)
4. Liu, M., Ben, L.: Research on demand forecasting method of multi-user group based on big data. In: Yamamoto, S., Mori, H. (eds.) Human Interface and the Management of Information: Applications in Complex Technological Environments. HCII 2022. LNCS, vol. 13306, pp. 45–64. Springer, Cham (2022). https://doi.org/10.1007/978-3-031-06509-5_4
5. Cong, Y., Yu, S., Chu, J., Su, Z., Huang, Y., Li, F.: A small sample data-driven method: User needs elicitation from online reviews in new product iteration. Adv. Eng. Inform. **56**, 101953 (2023)
6. Jin, J., Jia, D., Chen, K.: Mining online reviews with a Kansei-integrated Kano model for innovative product design. Int. J. Prod. Res. **60**(22), 6708–6727 (2022)
7. OPEN HowNet Git Code Homepage. https://gitcode.com/thunlp/openhownet/blob/master/README.md
8. OPEN HowNet Download. https://openhownet.thunlp.org/downloadThe
9. Qi, F., Yang, C., Liu, Z., Dong, Q., Sun, M., Dong, Z.: OpenHowNet: An Open Sememe-based Lexical Knowledge Base. arXiv preprint arXiv:1901.09957 (2019)
10. Dong, Z., Dong, Q.: HowNet-a hybrid language and knowledge resource. In: Proceedings of NLP-KE (2003)
11. Bello, A., Ng, S.-C., Leung, M.-F.: A BERT framework to sentiment analysis of tweets. Sensors **23**, 506 (2023)
12. Nagamachi, M.: Kansei engineering: a new ergonomic consumer-oriented technology for product development. Int. J. Ind. Ergon. **15**(1), 3–11 (1995)
13. Osgood, C.E., Suci, G.J., Tannenbaum, P.H.: The measurement of meaning (No. 47). University of Illinois Press (1957)
14. Shahapure, K.R., Nicholas, C.: Cluster quality analysis using silhouette score. In: 2020 IEEE 7th International Conference on Data Science and Advanced Analytics (DSAA), pp. 747–748. IEEE (2020)
15. Zhu, Y.L., Min, J., Zhou, Y.Q., Huang, X.J., Wu, L.: Semantic orientation computing based on HowNet. J. Chin. Inf. Process. **20**(1), 14–20 (2006)
16. Liu, J., Xu, J., Zhang, Y.: An approach of hybrid hierarchical structure for word similarity computing by HowNet. In: Proceedings of the Sixth International Joint Conference on Natural Language Processing, pp. 927–931 (2013)
17. Qiu, G.-P., Yuan, J.-L., Chen, J., Lin, M.-C.: Using kansei semantic differential method and grey relational analysis in the exploration of product form image cognition representation. In: Shin, C.S., Di Bucchianico, G., Fukuda, S., Ghim, Y.-G., Montagna, G., Carvalho, C. (eds.) AHFE 2021. LNNS, vol. 260, pp. 844–851. Springer, Cham (2021). https://doi.org/10.1007/978-3-030-80829-7_103
18. Zhang, N., Qin, L., Yu, P., Gao, W., Li, Y.: Grey-Markov model of user demands prediction based on online reviews. J. Eng. Des. **34**(7), 487–521 (2023)

Research on Intelligent Cabin Design of Camper Vehicle Based on Kano Model and Generative AI

Miao Liu, Zeming Zhao(✉), and Bo Qi

East China University of Science and Technology, Shanghai 200237, People's Republic of China
15954391269@163.com

Abstract. This study analyzes the functional requirements of different types of camping users in China based on the Kano model and improves the user experience of intelligent cabins using generative AI technology. The research adopts a combination of questionnaire surveys, literature review, and user interviews to construct a Kano model evaluation system. The Better-Worse coefficient method is used to analyze different functional attributes and identify priority sequences. Simultaneously, a detailed analysis is conducted on how generative AI affects interaction modes and interface design. The results reveal that cabin environment adjustment and entertainment functions are basic attributes, while voice navigation and intelligent voice assistants are expected attributes. Information push, intelligent recommendation, emotion recognition, and personalized experience are attractive attributes. Generative AI enables smarter voice interactions and more interesting personalized interfaces. Additionally, this study discusses smart cabin design strategies for camping vehicles based on the differences in functional priority among different user groups. Meeting basic needs is crucial, while customization of software and content design should be emphasized. The introduction of generative AI technology will drive interaction towards personalization, better satisfying diverse user needs.

Keywords: Intelligent vehicle cabin · Generative AI · User experience design

1 Background

In November 2022, the "Guiding Opinions on Promoting the Healthy and Orderly Development of Camping Tourism Leisure «pointed out that "camping tourism leisure refers to activities conducted outdoors using self-provided or leased equipment for leisure, sports, recreation, and nature education, primarily residing in designated camping areas with corresponding facilities [1]." Iimedia Consulting data show that the core market size of China's camping economy will rise to 248.32 billion yuan in 2025 [2].

The increase in disposable income of Chinese consumers is an important factor, and the upgrading of consumption structure makes camping enthusiasts' camping needs become more personalized and diversified, and the new consumption scene "camping + " model of camping 2.0 continues to develop.

H. Krömker (Ed.): HCII 2024, LNCS 14733, pp. 156–168, 2024.
https://doi.org/10.1007/978-3-031-60480-5_9

1.1 Camping Vehicle

Camper car is a relative concept, in this article refers to the car used for camping behavior. According to the different camping methods, can be divided into conventional camping, RV camping, car camping, special camping [3].

From the Perspective of Usage Scenarios. Conventional camping vehicles are similar to self-driving tour vehicles. RVs integrate the functions of mobile homes. Automobile camping vehicles emphasize loading capacity. Specialized camping vehicles are designed for specific activities, such as cliff camping.

From the Perspective of Development Both Domestically and Internationally. Camping industries in Europe and America have matured, while China is still in its initial stages. Due to the lack of infrastructure and supporting networks for RVs, Chinese vehicle owners often choose SUVs or MPVs that meet their daily driving needs for modification. These types of vehicles have ample space and can be equipped with tables, projection devices, and other amenities [4]. Some owners also opt to modify large space vehicles such as mini trucks or pickups.

From the Perspective of Smart Cabin Requirements. Pure wilderness camping emphasizes environmental adaptability, while camping at designated sites focuses on convenience, and short-term camping emphasizes practicality. Currently, RVs in the Chinese market are priced high and have low daily usage frequency. Therefore, the development of "camping mode" intelligent cockpit targeting SUVs, such as those developed by Tesla, NIO, and Xiaopeng, will expand the market capacity.

1.2 Intelligent Cockpit

The intelligent cockpit refers to the space within a vehicle that is relevant to human occupants, where the cabin and its associated functionalities are capable of autonomously sensing the status of passengers and the surrounding vehicle environment, making intelligent judgments and providing feedback. Through multimodal three-dimensional interaction, it achieves collaborative driving between humans and vehicles and, combined with user preferences, offers personalized services to each vehicle occupant [5].

The key technologies of intelligent cabin mainly include mechanical technology, electronic hardware technology, software technology and auxiliary information technology. Among them, artificial intelligence technology and cloud computing technology are considered to be the two major technological cornerstones [6].

From the Perspective of Key Technologies. Mechanical technology primarily optimizes internal space through mechanisms such as folding and sliding, including features like folding seats, movable bed-table-chairs, and adjustable roofs. Electronic technology encompasses sensor data collection, display technology for human-machine interaction, and electronic control systems for equipment control. In comparison to regular automobiles, sensors in smart systems of camping vehicles prioritize enhancing quality of life and usage safety. Different users have varying screen requirements. In the future, multi-screen interaction methods will integrate central information display, digital dashboards, heads-up displays, and rear-seat displays [7]. Control panel design influences human-machine experiential perception. Software technology addresses user experience through in-vehicle software, enabling remote smart interaction via mobile phones.

From the Perspective of Technological Evolution. Artificial intelligence technology provides the core intelligence for perception, analysis, decision-making, and interaction in intelligent cockpit. Meanwhile, cloud computing technology offers elastic, scalable storage and computing resources.

2 Generative AI

Generative artificial intelligence refers to the artificial intelligence technology that uses generative models to independently generate new data or content [8]. Generative AI models can be roughly divided into two categories: single-modal model and multi-modal model [**Σφάλμα! Το αρχείο προέλευσης της αναφοράς δεν βρέθηκε.**]. The integration of generative AI into the intelligent cockpit is characterized by autonomy, adaptability, continuity and interactivity.

2.1 Impact of Generative AI on Interaction

The collaboration between Mercedes-Benz and Microsoft applying ChatGPT in the automotive industry indicates continuous evolution in the core interaction modes of intelligent cockpit. The application of generative AI based on large-scale language models facilitates the integration of various modes into unified voice interactions.

In Terms of Interaction Modes. Current voice assistant functionalities are limited to preset tasks. Large pre-trained language models connect sensors, collect diverse user interaction data, deeply understand intent, respond in voice, adjust dynamically, and optimize training via user logs [10].

In Terms of Interaction Content. Vehicle-Generated-Content (VGC) emerges in automotive production [11]. In-vehicle systems adopt camping roles, offering comprehensive services. The intelligent voice assistant enables voice control, customizes services, engages in reasoning, and provides tailored responses or assistance.

In Terms of Experience Enhancement. Experience enhancement involves emotion recognition, environmental adjustment, personalized content generation using multi-source data, and proactive travel planning based on user profiles and data analysis.

The Impact of Generative AI on Interface Design. Generative AI offers new potential for personalized inter.face design in intelligent cockpits. It outputs text, images, and content for enhanced visual experience. Larger central control screens and integrated designs support this advancement, as seen in the Galaxy NOS system in the Geely Galaxy L6, which offers personalized desktop decoration through AI-generated WOW wallpaper. Personalized information streams are recommended based on user interests, boosting efficiency. Customized interactive content and entertainment experiences for children are feasible for family users.

3 Research and Results

3.1 Kano Model

The Kano model, invented by Professor Noriaki Kano [12]. The Kano model is illustrated in Fig. 1.

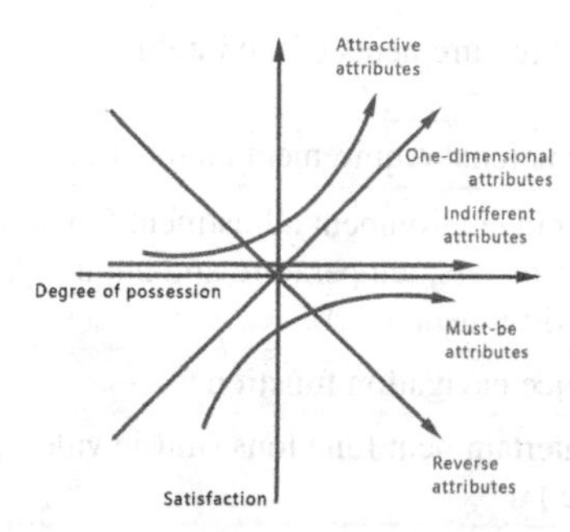

Fig. 1. Kano model

The Kano model categorizes user needs into five categories. In this study, the Likert scale was used to test the target users. Comparison of Kano rating classification results can be consulted according to Table 1.

Table 1. Comparison of Kano rating classification

Function		Negative items				
		Dislike	Can tolerate	Indifferent	Take for granted	Like
Positive items	Dislike	Q	R	R	R	R
	Can tolerate	M	I	I	I	R
	Indifferent	M	I	I	I	R
	Take for granted	M	I	I	I	R
	Like	O	A	A	A	Q

Berger et al. [13] proposed the use of the Better-Worse coefficient to calculate the percentage of functional requirements attributes classified. The specific formula for calculating the Better-Worse coefficient is as follows:

$$\text{Better} = (A + O)/(A + O + M + I) \tag{1}$$

$$\text{Worse} = -1 \times (O + M)/(A + O + M + I) \tag{2}$$

3.2 Surveys and Questionnaires

This paper utilizes literature review, interviews, and user surveys to gather opinions on basic and generative AI functions of intelligent cockpits, extracting user needs through affinity diagramming for classification and consolidation. Design attributes of the intelligent cockpit are determined accordingly, detailed in Table 2.

Table 2. The functional requirements of the intelligent cabin of the camping car

Category	Number	Functional requirement element
Basic function	D1	Cabin environment adjustment function (multi-functional seat parameters, temperature and humidity, sound and light environment, etc.)
	D2	Voice navigation function
	D3	Entertainment functions (music videos, KTV mode, cinema mode, etc.)
	D4	Camping information push function (push camping site, surrounding attractions, restaurants and hotels on the screen)
Generative AI function	S1	Intelligent Voice Assistant (incorporating multimodal fused generative AI functionality)
	S2	Emotion recognition and personalized experience (personalized modeling based on information such as sensors and user records, actively adjusting seats, switching air conditioners, playing music, etc.)
	S3	Intelligent recommendation system (according to user preferences to actively recommend natural landscapes, customs, camping sites, etc.)
	S4	Children's entertainment functions (children's picture books combined with generative AI, children's songs, etc.)

The Kano questionnaire is tailored to the functional needs of users regarding Intelligent cockpit. Target users include family campers, couples, self-driving tourists, veteran campers, and retirees. Surveys were conducted at campsites and online, yielding 392 valid responses. Specific respondent demographics are detailed in Table 3.

Table 3. Composition of respondents

Classification	Family campers	Couple travel group	Self-driving tourist	Veteran camper	Retirement camp
Frequency	98	69	89	81	55
Proportion	25.00%	17.60%	22.70%	20.66%	14.03%

3.3 Result

Cronbach's α coefficient assessed questionnaire reliability, yielding coefficients ranging from 0.7 to 0.9, indicating good reliability [14]. Functional requirements of Intelligent cockpit were categorized using the Kano model, with Better and Worse coefficients calculated via Eqs. (1) and (2). Classification results are summarized in Table 4.

Table 4. Classification of Demand Factors Properties

Function	A	O	M	I	R	Q	Result	Better	Worse
D1	19.69%	29.41%	**35.04%**	15.09%	0.77%	0.00%	M	49.48%	−64.95%
D2	20.46%	**31.97%**	27.62%	18.93%	0.51%	0.51%	O	52.97%	−60.21%
D3	22.51%	28.13%	**30.95%**	17.39%	1.02%	0.00%	M	51.16%	−59.69%
D4	**42.97%**	16.62%	16.11%	22.51%	0.51%	1.28%	A	60.68%	−33.33%
S1	30.69%	**32.48%**	11.00%	24.81%	0.77%	0.26%	O	63.82%	−43.93%
S2	**37.08%**	35.81%	9.21%	17.14%	0.77%	0.00%	A	73.45%	−45.36%
S3	**39.39%**	34.78%	12.02%	13.55%	0.26%	0.00%	A	74.36%	−46.92%
S4	32.48%	12.28%	8.95%	**44.76%**	1.02%	0.51%	I	45.45%	−21.56%

A Better-Worse coefficient quadrant chart is plotted, as shown in Fig. 2.

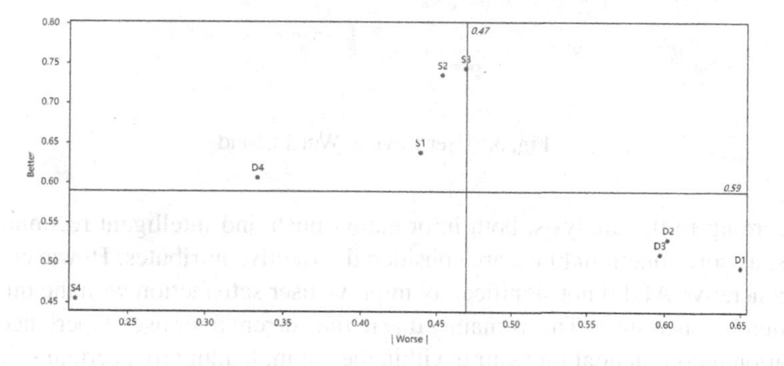

Fig. 2. Better-Worse coefficient quadrant

The environmental regulation functions and entertainment features within the cabin, such as seat parameter adjustments, air conditioning temperature control, music, and video, represent the most basic requirements for campers regarding onboard facilities under contemporary technological conditions. If these functions cannot be fulfilled or directly affect their usage experience and purchasing decisions, they must be prioritized for resource allocation to ensure enhancement, serving as essential attributes of the product.

Voice navigation and intelligent voice assistant functions, while users may not actively or directly request such demands, providing relevant services proactively by the system can surprise users and enhance satisfaction. These functions provide assistance from the dimensions of assisted driving and tourism, significantly improving user impressions afterward, thus requiring emphasis as one-dimensional attributes.

As for personalized experiences, related information push notifications, intelligent recommendations, and other functions, ordinary users may find it challenging to conceive or demand such innovative services. Still, once experienced, they can bring extraordinary personalized travel enjoyment. Therefore, product designers should actively provide these charm attributes to differentiate the product.

Finally, children's entertainment functionalities have relatively limited usage scenarios and should not occupy too much proportion in resource allocation. Selective inclusion of certain games, stories, and other content based on vehicle type and target users is advisable, but they need not be considered as key attributes to be taken into comprehensive consideration.

Fig. 3. User Review Word Cloud

According to the analysis, both information push and intelligent recommendation features, as core functionalities, are considered attractive attributes. However, incorporating generative AI did not significantly improve user satisfaction with the information recommendation system. This is mainly due to the current poor user experience with the information recommendation feature within the cabin, leading to a certain level of user confidence deficiency in this functionality aspect, as depicted in Cloud Fig. 3.

Intelligent recommendation systems should deeply integrate generative models, data, users, and the environment to achieve intelligent and personalized user experiences. This involves connecting with professional vertical domain sources for authoritative recommendations, refining user profiles for customized recommendations, optimizing pages and interactions for enhanced user experience, and dynamically iterating updates and incorporating real-time external data for optimal solutions. Development processes should focus on the user's entire lifecycle and seamless integration with the mobile end. Overall, achieving true intelligent interaction and a good user experience in intelligent recommendation systems necessitates support from data, user customization, environmental matching, and seamless integration. To enhance these functions, specialized large models tailored for camping scenarios should be employed, offering greater professionalism and practicality, and providing more efficient and intelligent services [15].

Table 5. The impact of user experience on results

Topic	Option	Have used generative AI before		Aggregate
		Yes	No	
Understand personalized experience and intelligent recommendation	Yes	45(78.95%)	12(27.91%)	57(57.00%)
	No	2(3.51%)	7(16.28%)	9(9.00%)
	Inadequacy	9(15.79%)	24(55.81%)	33(33.00%)
	Else	1(1.75%)	0(0.00%)	1(1.00%)
Aggregate		57	43	100

The application of generative AI in intelligent cockpit is still in its nascent stage, with users largely lacking relevant experiences. Surveys indicate that nearly sixty percent of users have utilized generative AI, with this demographic demonstrating a better understanding of the value of personalized experiences and intelligent recommendations, as shown in Table 5. However, among non-users, over half cannot envision its application scenarios and experiential effects. Overall, generative AI-related functionalities are currently positioned as attractive attributes, but this is due to users' limited understanding of its application scenarios and effects. As user experiences become more enriched, such functionalities will become increasingly significant.

Table 6. User interest in generative AI

Topic	Option	Have used generative AI before		Aggregate
		Yes	No	
Trying out the AI-powered image generation feature on Geely cars	Interesting	48(84.21%)	23(53.49%)	71(71.00%)
	Feeling of no practical use	12(21.05%)	12(21.05%)	25(25.00%)
	Incomprehension	8(14.04%)	18(41.86%)	26(26.00%)
Aggregate		57	43	100

The sampling survey indicates, as shown in Table 6, that nearly eighty-five percent of users who have experienced generative AI express interest in the AI-generated image function. Among non-users, over half still express interest in this function. This suggests that the integration of cutting-edge technologies such as generative AI, including AI-generated images, can effectively stimulate consumers' interest and attention towards intelligent cockpit overall. This is primarily attributed to the anticipation of novelty in the application of new technologies and the imagination of potential new experiences they may bring. Therefore, actively utilizing cutting-edge technologies such as generative AI for product and interface innovation will help intelligent cockpit better attract target users, thus achieving effective market promotion.

3.4 User Group Segmentation

Based on KANO questionnaire results, camping demographics prioritize smart cabin features differently. Hence, this study creates Better-Worse coefficient quadrant diagrams separately for various user groups to propose strategies for improving and optimizing smart cabin designs for camping vehicles, as shown in Fig. 4.

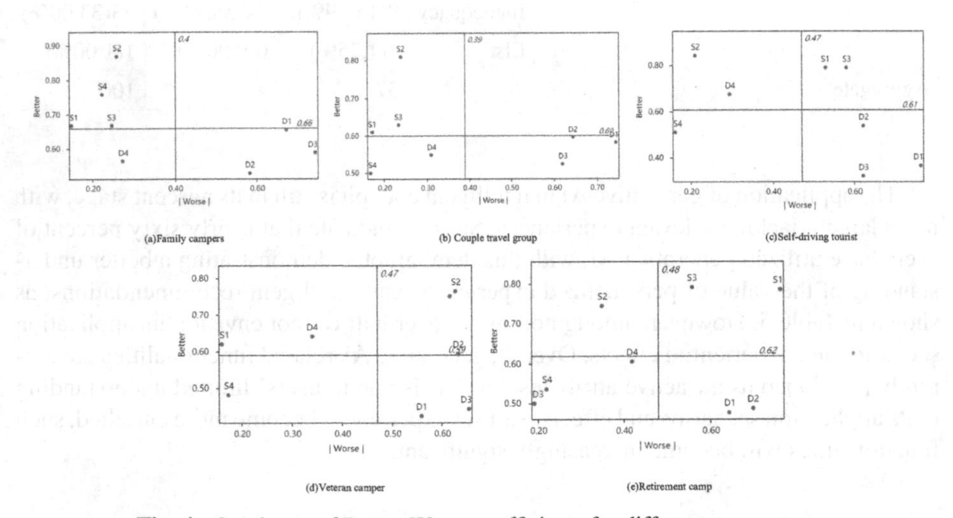

Fig. 4. Quadrants of Better-Worse coefficients for different user groups

Family Camping Group. This demographic primarily consists of couples camping with children or elderly relatives, focusing on the enhancement of family relationships during camping. They often choose campsites with well-equipped facilities or convenient locations such as urban outskirts for relaxed and leisurely camping. They value entertainment features and information recommendations provided by intelligent cabins to help create a relaxed and harmonious travel experience.

Customized Camping Information Recommendations Based on Family Member Preferences. For example, pushing scenic details to the elderly and recommending campsite selections with amusement facilities for children to meet diverse needs.

Interactive Learning and Entertainment Applications for Family Members. During the journey, intelligent cabins provide children's entertainment systems, interactive learning games, etc., to combine education with entertainment and enrich the travel experience. More intelligent management of family member needs through entertainment features such as music and videos meets the demands of the group.

Couples Traveling Group. This demographic emphasizes experiential designs focusing on intimacy and ambiance creation. Many young couples prefer more refined camping methods, focusing on comfort and experiential aspects during the process.

Customized Information Recommendations Based on Couple Preferences. The system can push destination recommendations and support itinerary planning based on the interests of the couple, forming customized solutions that align with both parties' interests. For couples who enjoy traveling with pets, recommending places that allow animals is particularly important.

Digitally Creating an Ideal Romantic Camping Atmosphere. Supporting couples in selecting lighting, sound effects, and other parameters based on their interests to create a unique ambiance. Many couples believe that various personalized modes greatly enhance satisfaction with the experience.

Self-driving Tourists. This group typically engages in multi-destination camping trips and emphasizes intelligent cabins to meet customized needs, achieving flexible travel.

Personalized Route and Dynamic Attraction Recommendations. Real-time updates on niche route information, especially introductions to natural scenic spots, inspire proactive exploration of surrounding possibilities. Dynamic adjustments to recommendations based on real-time traffic and weather guide travelers to roam freely.

Intelligent Voice Navigation and Real-Time Traffic Status Updates. Real-time updates on navigation and RV resource distribution, along with refined prompts, avoid detours. Connecting with surrounding vehicles and infrastructure ensures more timely and reliable information.

Itinerary Management and Video Communication Tools Assistance. Campers organizing self-driving tours can utilize features like schedule planning to reach consensus on itineraries in advance, then import them into the system. Video call functions make real-time communication during caravan travel more convenient.

Seasoned Camping Enthusiasts. This demographic pursues a primitive adventure experience, opting for deep travel in remote areas.

Recommendation Planning Function Focusing on Exploration and Risk Avoidance. The system can preemptively push destination-related natural geography and meteorological information, particularly in areas with significant cultural differences, analyzing and prompting relevant taboos to reduce potential cultural clashes. Real-time route recommendations also inspire deeper exploration of nature.

Intelligent Environment Control Ensuring Travel Comfort. Based on user and environmental data, the system achieves personalized optimization of parameters such as seats, temperature, and sound effects to ensure cabin comfort. Controlling various parameters through the "camping mode" creates a relaxing environment through environmental creation.

Safety Monitoring and Warning Function Ensuring Risk Avoidance: -time monitoring of driver status and alerts regarding fatigue driving risks; detection of surrounding environments during overnight stays in remote areas provides peace of mind.

External Drone Photography Supporting Digital Recording of Exploration Experiences. Meets the needs for recording and photography.

Retired Camping Community. This group's travel focuses more on leisure and entertainment, but with limitations in physical stamina and safety. The group values cabin comfort, personalized health management, and intelligent voice interaction.

Comfortable Cabin Environment Considering Physical Condition. Based on age-appropriate data models, parameters such as seats, temperature, and lighting are automatically optimized to create a relaxed space. Massage functions help relax and alleviate travel fatigue.

Personalized Health Plan and Reminder Services. The system timely reminds of medical consultations, medication, etc. Intelligent technology achieves precise modeling and management of travel health for the elderly, supporting their peace of mind during travel.

Intelligent Voice Interaction and Translation Reducing Technical Barriers. Voice assistants allow the elderly to query resources and manage equipment through voice commands, reducing learning costs. Recognizing different languages for real-time translation overcomes language barriers.

4 Discussion

Based on the above analysis, the design strategy for intelligent cockpit functions can be summarized as follows:

Standard Configuration Ensure. Basic needs are met, including fundamental functions like environmental regulation (temperature, humidity, ventilation), catering to the common needs of most users.

Personalized Software. Develop models tailored to different camping scenarios and user groups (e.g., families, couples), offering personalized downloads such as camping mode and KTV mode.

Enhanced Vehicle-Machine Interconnection. Improve interoperability between vehicles and mobile phones for seamless function and experience integration, enhancing satisfaction for camping groups utilizing mobile phones.

Flexible Customization. Offer seat customization [16], screen selection combinations, and other personalized solutions for important functions like seats and screens, enhancing emotion recognition and personalized experiences, particularly in professional camping vehicles.

Intelligent Interaction. Develop expected intelligent interaction functions based on generative AI, such as voice assistants and intelligent information recommendation systems, focusing on relevance and simplicity enhancements.

Exploration of Frontier Technologies. Exploit frontier technologies like generative AI to introduce novel features such as personalized wallpapers, arousing user interest despite current interface design immaturity.

5 Result

On the basis of user research, this paper uses Kano model to analyze the preferences of different types of campers for intelligent cabin functions, and mainly draws the following conclusions: First, basic functions are still the basic needs of users. Environmental regulation, entertainment functions, etc., are the basic attributes of the car, which must be met first. Second, expected functions such as voice and recommendation deserve priority. Navigation and voice assistants provide additional satisfaction and are a product differentiator. Third, the charm functions such as emotion recognition and personalized content have innovative potential. This kind of cutting-edge demand can break through the traditional car application category, and show the brand's scientific and technological strength while satisfying users. Fourth, different camping groups have different functional needs. Design solutions for specific groups to achieve differentiated competitive advantages. This study also has some shortcomings. First, the sample size is limited and cannot fully represent all potential users. Secondly, the evaluation of some novel functions may be biased. Follow-up studies will be conducted with a large sample of a broader user base and with more rigorous evaluation methods. In addition, this study only stays at the user demand level, and it is necessary to deeply study the specific application scenarios of generative AI in intelligent cockpit in future design practices, carry out in-depth user research, and truly realize human-oriented technology application and scenario design.

References

1. Central People's Government of the People's Republic of China. https://www.gov.cn/gongbao/content/2022/content_5674298.htm, (Acessed 25 Jan 2024)
2. Iimedia Homepage. https://www.iimedia.cn/c400/95478.html, (Accessed 25 Nov 2023)
3. Li, Z., Du, X., Shen, H.: Development history and prospect of camping tourism in China. China Tourism News (3) (2023)
4. Zhao, Y., Liu, J., Lei, C., et al.: Intelligent cabin design based on MPV model. Manufact. Autom. **45**(08), 164–166 (2023)
5. Cai, M., Wang, W.: A review of research on interactive design of intelligent car cabins. Packaging Eng. **44**(06), 430–440 (2023)
6. Du, Z., Huang, X., Meng, J.: Key technologies of intelligent cabins. Era Automobiles (05), 143–144 (2021)
7. Wang, B.: The Past, Present, and Future of Intelligent Cabins (07), 167–168 (2022)
8. Gartner Homepage. https://www.gartner.com/en/articles/5-impactful-technologies-from-the-gartner-emerging-technologies-and-trends-impact-radar-for-2022, (Accessed 24 Nov 2023)
9. Cao, Y., et al.: A comprehensive survey of ai-generated content (aigc): A history of generative ai from gan to chatgpt. (2023) arXiv preprint arXiv:2303.04226
10. Xiao, Y.: Generative language models and general artificial intelligence: connotation, paths, and implications. People's Forum: Academic Front. (14), 49–57 (2023)
11. Qin, J., He, J.: Research on scene interaction design of metaverse intelligent cabins for autonomous vehicles. Packaging Eng. **44**(18), 67–76 (2023)
12. Kano: Attractive quality and must-be quality. J. Japanese Soc. Qual. Control **14**(2), 147–156 (1984)
13. Berger, C., Blauth, R., Boger, D., et al.: Kano's methods for understanding coustomer-defined quality. Center Qual. Manag. J. **2**(4), 3–36 (1993)

14. Eisinga, R., Grotenhuis, M., Pelzer, B.: The reliability of a two-item scale: Pearson, Cronbach, or Spearman-Brown? Int. J. Public Health **58**, 637–642 (2023)
15. Zhao, Z., Wang, Z.: Overview and trend analysis of domestic large-scale models development in 2023. Young J., 1–4 https://doi.org/10.15997/j.cnki.qnjz.20231128.001
16. Liu, J., Lu, Y., Xu, Z., et al.: Empirical study on sleep comfort of intelligent cockpit. Packag. Eng. **41**(22), 66–71+82 (2020)

A Study of the Effects of Different Semantic Distance Icons on Drivers' Cognitive Load in Automotive Human-Machine Interface

Yanchi Liu, Wanrong Han, Dihui Chu, and Jianrun Zhang(✉)

School of Mechanical Engineering, Southeast University, Nanjing 211189, China
220214994@seu.edu.cn

Abstract. This study investigates the impact of different semantic distances of icons in the automotive human-machine interface (HMI) on drivers' cognitive load. Experiments were conducted with 20 experienced drivers to assess their response time, accuracy rate, and subjective cognitive load measured by the NASA-TLX scale while identifying icons with varying semantic distances. The results showed that icons with a greater semantic distance increased the drivers' response time and cognitive load, and decreased recognition accuracy. These findings are significant for optimizing automotive HMI design, guiding designers to consider the intuitiveness and recognizability of icons to reduce drivers' cognitive load and enhance road safety. This research provides an empirical foundation for improving the driving experience and promoting more human-centered automotive interface design.

Keywords: Automotive Human-machine Interface · Icon Design · Semantic distance · Cognitive load

1 Introduction

With the continuous development of technology, the increasing complexity of modern car's infotainment systems and control interfaces has raised concerns about drivers' cognitive load. Drivers need to process information from multiple sources while driving, including but not limited to road traffic conditions, navigation instructions, entertainment content, and communication information. The parallel execution of these tasks can significantly increase drivers' cognitive load, thereby weakening their perception and reaction speed to road traffic conditions and increasing driving risks [1, 2]. In this context, cognitive load is defined as the psychological and cognitive stress experienced by users when performing specific tasks, directly affecting their decision-making ability and task performance efficiency [3]. Research by Engström (2017) and others has shown that in complex driving environments, the level of cognitive load of drivers is closely related to the risk of traffic accidents. Drivers are prone to distraction when handling high cognitive load tasks, thereby increasing the likelihood of accidents [4]. Therefore, studying drivers' cognitive load and finding effective design solutions to reduce it is of great importance to improve road safety.

H. Krömker (Ed.): HCII 2024, LNCS 14733, pp. 169–182, 2024.
https://doi.org/10.1007/978-3-031-60480-5_10

Icons in the automotive human-machine interface, as visual elements that convey functions and operations, directly impact drivers' experience and cognitive load. Research by McDougall and de Bruijn (2000) and others found that the choice of icons affects users' perception and performance, with different types of icons differing in information conveyance [5]. This study focuses on the semantic distance of icons, i.e., the closeness of the icon itself to the function it represents. Some icons directly convey functional meanings, allowing users to quickly recognize their operational meanings, which means the icons have a closer semantic distance. In contrast, other icons may require more cognitive resources to understand their represented functions, indicating a greater semantic distance [6, 7].

Through empirical research on the relationship between different semantic distances of icons and drivers' cognitive load, this study aims to provide specific and targeted guidance for automotive human-machine interface (HMI) design. Using multiple indicators such as drivers' response time, accuracy rate, and subjective cognitive load, this study seeks to offer scientifically sound suggestions for automotive HMI design to effectively reduce drivers' cognitive load, improve their performance in multitasking operations, and thereby enhance driving safety and comfort. In this context, this paper will delve into the specific impact of different semantic distances of icons on drivers' cognitive load, aiming to make a substantial academic contribution to the field of automotive HMI design.

2 Methodology

2.1 Materials

The experimental materials were based on a survey of existing automotive HMI (Human-Machine Interface) function categories, selecting eight key functions in the automotive HMI: seat, setting, phone, sound volume, navigation, home, air conditioning, and music, to ensure representativeness and comprehensiveness of the materials. Under each function category, six icons were selected, totaling 48 icons. The style of the icons mainly embodied a two-dimensional and minimalist concept, and a consistent color scheme of white on black was maintained to avoid distractions caused by color contrast. Special attention was paid to ensure diversity in design and semantic distance, so that experts and subjects could visually distinguish each icon clearly while also providing a more diverse range of choices. The size of the icons was chosen to be 70*70px to ensure clarity and recognizability on screens of different sizes, with specific examples shown in Fig. 1.

To ensure a comprehensive and professional selection of icons with varying semantic distances, seven experts experienced in interface design were invited to independently rate the 48 selected icons. During the expert rating process, each expert used a rating scale from 1 to 5 for each icon, where 1 indicated the icon was unrelated to its function and 5 indicated the icon was completely related to its function, reflecting the degree of closeness of the icon to its corresponding function.

After collecting the expert rating data, systematic data processing was carried out to calculate the average score for each icon. Using the classification method proposed by Peirce in 1958, the icons were divided into three main categories: the first level directly representing the function, the second level metaphorically or indirectly representing the

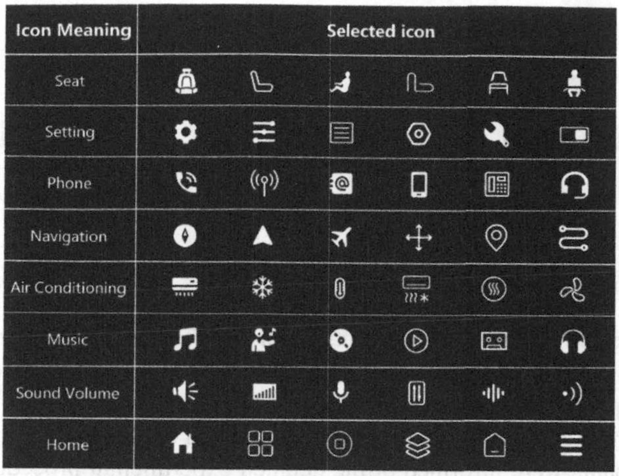

Icon Meaning	Selected icon
Seat	
Setting	
Phone	
Navigation	
Air Conditioning	
Music	
Sound Volume	
Home	

Fig. 1. Automotive HMI 8 function icon selection.

function, and the third level being unrelated to the function [8]. Following these steps, based on the results of the expert survey, icons for each level of functional meaning were finalized, resulting in a total of 24 icons, with specific examples shown in Fig. 2.

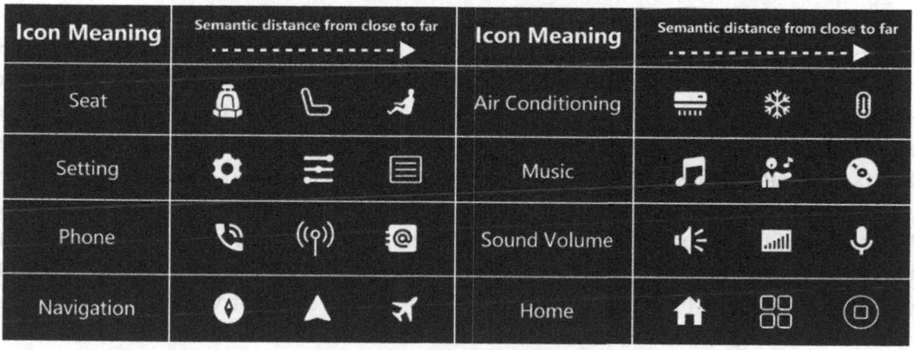

Icon Meaning	Semantic distance from close to far	Icon Meaning	Semantic distance from close to far
Seat		Air Conditioning	
Setting		Music	
Phone		Sound Volume	
Navigation		Home	

Fig. 2. Classification of icons with different semantic distances.

After obtaining the icons for the three categories, they were applied to three automotive HMIs as experimental materials. In the bottom bar of each automotive HMI, a group of 8 icons was arranged in order, maintaining consistency in background color, icon size, and arrangement. In terms of design, the other parts of the HMI utilized low-fidelity prototypes to minimize potential distractions. The aim of this design was to eliminate other potential factors besides the functional meaning of the icons, ensuring the reliability and comparability of the experimental results. The specific layout of the three automotive HMIs is shown in Fig. 3.

Fig. 3. Three automotive HMIs with different semantic distance icons.

2.2 Participants

This study selected 20 graduate students from Southeast University as participants, including 10 males and 10 females, with an age range of 21 to 26 years (average age = 24). All participants had normal or corrected-to-normal vision and were in good health. They all had driving experience and were familiar with automotive control systems.

All participants were informed of the detailed experimental procedures and operational requirements before the start of the experiment to ensure its smooth conduct. The experiment used a 16-inch LCD screen with a resolution of 2560 × 1600 pixels for participants to view the experimental materials. Each participant completed the experiment in a brightly lit, quiet room.

2.3 Experimental Procedure

The aim of this study is to evaluate the impact of icons with different semantic distances in automotive Human-Machine Interfaces (HMIs) on drivers' cognitive load. The experimental materials included the three HMIs discussed previously, each featuring a set of eight icons with varying semantic distances, arranged in the bottom bar of the automotive HMI. To ensure comparability and accuracy in the experiment, the three HMIs maintained consistency in background color, icon size, and arrangement.

This experiment included a practice phase and three successive experimental stages. Each stage of the experiment displayed a different automotive HMI, each with a bottom bar containing eight icons. The display time for each interface was set to 3000 ms, based on previous research by Ma et al. in 2015, which indicated that driver distraction lasting more than 3 s significantly increases the risk of accidents [9].

At the beginning of the formal experimental stage, subjects were required to focus on a red cross appearing on the screen for 1000 ms to ensure their attention was concentrated. Subsequently, subjects randomly received task instructions, requiring them to interact with icons on the automotive HMI using a mouse. They needed to click on the icon that matched the task instruction within the specified time. These tasks were presented in a random order, covering eight operations: adjusting volume, adjusting seats, making phone calls, opening navigation, clicking on the home screen, turning on air conditioning, playing music, and accessing settings. For example, if the task "Make a Phone Call" appeared, the subject had to click on the icon they believed represented "Phone" on the interface. If the time taken to recognize the icon exceeded 3 s, the attempt was recorded as an error, and the recognition time was capped at 3 s. The experiment was conducted using E-Prime software, which was used to record the subjects' response times and accuracy rates in the experiment. The specific experimental procedure is shown in Fig. 4.

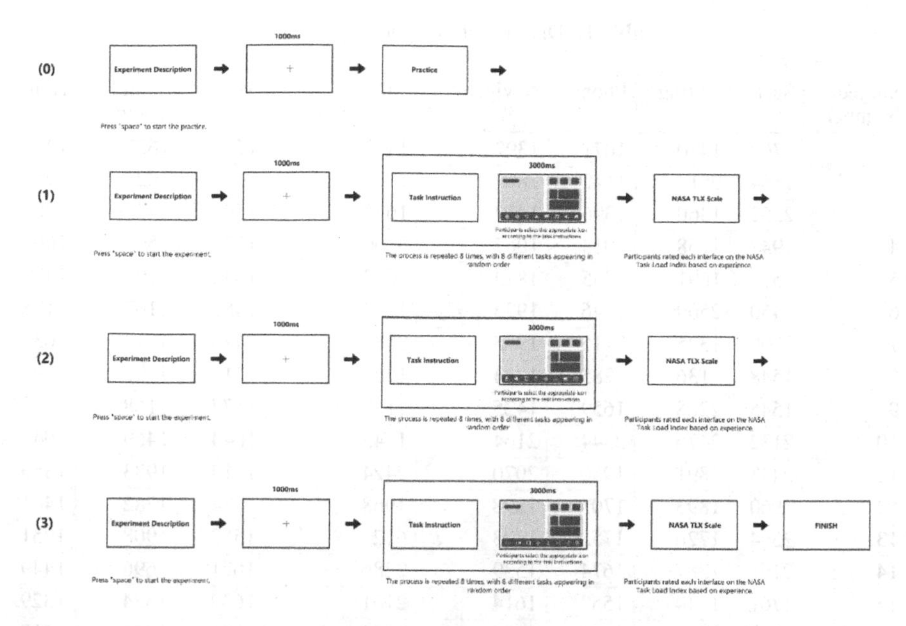

Fig. 4. Flowchart of the Experimental Procedure: (1) Practice Phase (2) First Set of Interface Experimental Phase (3) Second Set of Interface Experimental Phase (4) Third Set of Interface Experimental Phase.

3 Results

3.1 Task Response Time Data

This study quantified the specific impact of icon semantic distance on response time through three sets of experimental data. Each set of experiments corresponded to a set of icons with different semantic distances, namely the icons directly representing function meanings, icons metaphorically or indirectly representing function meanings, and icons unrelated to function, with semantic distances ranging from near too far.

In Table 1, the direct icons participants encountered were intuitively designed to reveal their function meanings, resulting in the shortest average response time (in milliseconds). Table 2 involved indirect icons, requiring participants to interpret the metaphors in the icon design, leading to relatively prolonged response times. In Table 3, the irrelevant icons faced by the participants had no direct association with their function meanings, leading to the longest recorded response times under this condition.

This study conducted a comprehensive analysis of the impact of three types of icons with different semantic distances on subjects' response times. The average response times for tasks involving direct, indirect, and irrelevant icons at various testing stages were derived from the three tables mentioned above, as shown in Table 4. These data clearly indicate a trend of increasing average response times for subjects as the semantic distance of the icons increases. Correspondingly, the line graph intuitively depicts the trend of response times as they vary with semantic distance, as shown in Fig. 5. Specifically, the overall average response time for direct icons was 1637 ms, for indirect icons

Table 1. Direct Icon Response Time Data.

Subject Number	Saet	Setting	Phone	Navigation	Air Conditioning	Music	Sound Volume	Home
1	1760	1340	1071	1392	1482	1704	1627	1241
2	1654	951	1142	784	1181	865	792	896
3	2252	1360	1396	1417	1676	1596	2427	1865
4	1948	1768	1013	1067	3000	1574	2595	1004
5	2577	1691	2235	1822	2032	1842	2102	1489
6	1450	2509	1746	1923	1382	1482	1167	1198
7	3000	1325	2715	1975	2885	1470	1944	1636
8	1548	2136	1285	1834	1220	1216	1208	1590
9	1545	2268	1659	1425	1139	1177	1168	1860
10	2132	2476	2344	2164	1543	1644	1469	1049
11	2433	1891	1950	2070	2424	1314	1933	1356
12	1960	1893	1703	1898	1968	1474	1782	1441
13	2364	1726	1753	1393	612	1552	1908	1251
14	2161	1387	1674	1539	2286	1671	1696	1449
15	1761	1244	1557	1614	2161	1632	1534	1329
16	1721	1396	1154	2006	1443	1393	1275	1520
17	1578	1724	1395	1664	1443	1284	1641	1458
18	1952	1864	1472	1500	1376	1404	1406	1077
19	1427	1673	1177	1681	1279	1464	1867	1405
20	2274	1447	1779	1430	1294	1372	1556	1392

Table 2. Indirect Icon Response Time Data.

Subject Number	Saet	Setting	Phone	Navigation	Air Conditioning	Music	Sound Volume	Home
1	1830	1709	2474	2404	2296	2041	2981	2361
2	1388	1006	3000	1962	1248	1498	1083	936
3	1998	1164	1659	2360	3000	3000	2483	1105
4	2613	1115	2906	1153	3000	1160	2945	1034
5	2415	1972	2214	3000	1526	2216	1752	1703
6	2600	1318	1757	2429	1957	1834	1648	1387
7	3000	2574	1998	3000	2294	1619	1575	1385
8	3000	1547	2037	1748	2458	2042	2175	3000
9	2488	1937	3000	1840	2648	2393	1514	2136
10	2142	2025	2243	2671	2449	1782	2839	3000
11	2161	1650	1867	2340	3000	1133	2856	2443
12	1461	1653	1554	2045	1785	1545	2666	3000

(*continued*)

Table 2. (*continued*)

Subject Number	Saet	Setting	Phone	Navigation	Air Conditioning	Music	Sound Volume	Home
13	2060	1423	3000	2179	2542	2745	2824	2211
14	2351	2954	1517	3000	2074	2051	2558	2441
15	1167	2757	1369	1558	1961	1950	1354	2186
16	1506	2967	2860	2231	2308	2335	2030	1391
17	3000	1420	1164	1645	2308	1857	2989	2160
18	1524	1613	1262	2363	1247	1365	2195	2152
19	1566	1349	888	1673	2159	1519	1773	2148
20	1927	1036	1650	2243	3000	2282	2383	3000

Table 3. Irrelevant Icon Response Time Data.

Subject Number	Saet	Setting	Phone	Navigation	Air Conditioning	Music	Sound Volume	Home
1	3000	2574	2545	3000	2683	2576	1770	1571
2	1279	1079	1834	1644	1551	2950	1641	1184
3	2021	1261	3000	1375	2498	3000	1375	1149
4	1967	1440	2093	1980	1733	2237	2311	1202
5	3000	2229	3000	2336	2309	3000	2331	1144
6	1949	2089	1515	2038	2663	3000	1993	2934
7	3000	3000	2411	2357	1962	2628	1456	2966
8	1765	1031	2931	2620	2903	2474	3000	2229
9	2223	3000	1573	3000	2350	3000	2245	1648
10	3000	2512	3000	2242	3000	2661	2383	3000
11	3000	2500	2789	3000	3000	3000	2775	2224
12	2248	2505	2372	3000	2680	2360	2521	2766
13	3000	2197	2456	2872	1023	2626	2732	1928
14	3000	1572	2322	2206	3000	3000	2377	2489
15	1856	1309	2126	2378	2915	2901	2106	2148
16	1775	3000	1446	3000	2045	2080	1673	2688
17	1493	2193	3000	2493	2044	3000	2285	2514
18	2333	2451	1982	2118	3000	2120	1893	1436
19	2388	2099	3000	2531	1846	2325	2664	2364
20	2869	1682	2501	1957	3000	2010	2144	2128

it was 2069 ms, and for irrelevant icons, it was 2322 ms. The data reveal an upward trend in the average response time of subjects as the semantic distance of the icons increases.

Subsequently, a one-way ANOVA was conducted on the relationship between icons of different semantic distances and response times. The results showed significant differences in response times among icons with three different semantic distances ($F = 67.109$, $P = 0.000^{***} \leq 0.05$), as shown in Table 5. This indicates that semantic distance is a

Table 4. Average Response Times.

Icon Meaning	Direct Icon Task Mean Response Time	Indirect Icon Task Mean Response Time	Irrelevant Icon Task Mean Response Time
Saet	1975	2110	2359
Setting	1703	1759	2086
Phone	1611	2021	2395
Navigation	1630	2192	2407
Air Conditioning	1691	2263	2410
Music	1457	1918	2647
Sound Volume	1655	2231	2184
Home	1375	2059	2086

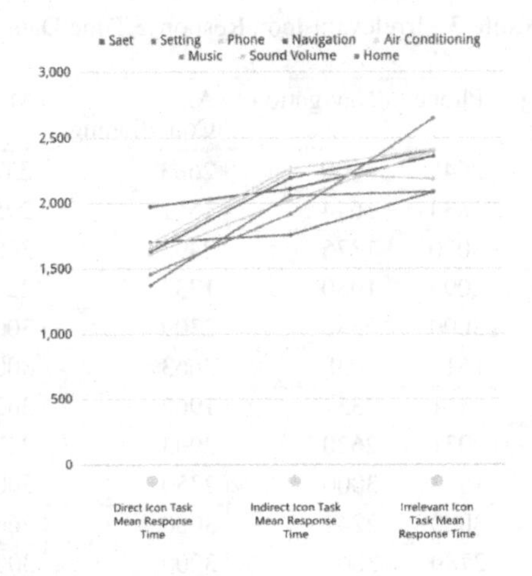

Fig. 5. Trends in task response times with icon semantics.

significant factor affecting the time users need to recognize an icon. Specifically, the average response time for direct icons was 1637.144 ms, for indirect icons it was 2069.238 ms, and the longest was for irrelevant icons, at 2321.7 ms. Welch's F test also yielded similar significant results (F = 79.661, P = 0.000*** ≤0.05), further confirming this trend. This statistical significance suggests that the time required for users to complete the recognition task increases with increasing semantic distance, indicating that considering the direct association between an icon and its functional meaning is one of the keys to enhancing user experience and efficiency in automotive HMI design.

Table 5. ANOVA Results for Response Time Across Icon Semantic Distances.

Dependent Variable	Source of Variation	sample size	Mean Squares	F-Value	Significance	Welch's F-Value
Response Time	Direct Icon	160	1637.144	435.614	F = 67.109 P = 0.000***	F = 79.661 P = 0.000***
	Indirect Icon	160	2069.238	596.534		
	Irrelevant Icon	160	2321.7	558.191		
	Total	480	2009.36	603.821		

3.2 Task Response Accuracy Rate Data

This section presents a set of tables and line graphs, recording the accuracy rates of 20 participants in the task of recognizing icons with three different semantic distances. The tables display the correct task response rates for direct, indirect, and irrelevant icons in the order of participant numbers, along with the average accuracy rates for each group of icon tasks, as seen in Table 6. These data reveal that participants generally exhibited higher accuracy in tasks involving direct icons, while accuracy was relatively lower in tasks with irrelevant icons.

Table 6. Correct task response rate.

Subject Number	Direct Icon Correct task response rate	Indirect Icon Correct task response rate	Irrelevant Icon Correct task response rate
1	100.00%	100.00%	87.50%
2	100.00%	87.50%	62.50%
3	87.50%	62.50%	62.50%
4	100.00%	87.50%	100.00%
5	100.00%	62.50%	75.00%
6	75.00%	87.50%	62.50%
7	87.50%	75.00%	75.00%
8	87.50%	50.00%	50.00%
9	100.00%	75.00%	37.50%
10	100.00%	75.00%	50.00%
11	87.50%	87.50%	50.00%
12	87.50%	75.00%	62.50%
13	75.00%	62.50%	62.50%
14	100.00%	87.50%	62.50%
15	100.00%	100.00%	87.50%
16	100.00%	62.50%	75.00%
17	100.00%	75.00%	50.00%
18	87.50%	75.00%	75.00%
19	100.00%	87.50%	62.50%
20	87.50%	62.50%	37.50%

The corresponding line graph draws individual lines for each participant's accuracy rate, differentiated by color coding, showing their performance in recognizing various types of icon tasks. The fluctuations in participants' accuracy rates in the line graph visually reflect the individual differences and overall trends in recognition accuracy across tasks with icons of different semantic distances, as detailed in Fig. 6.

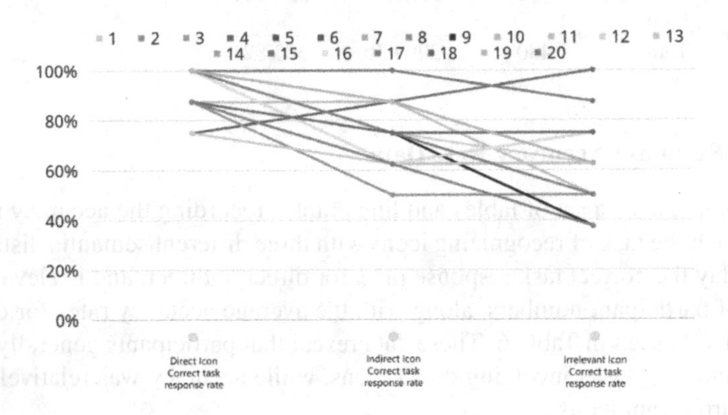

Fig. 6. Trends in Correct task response rate with icon semantics.

In the one-way ANOVA analysis examining the relationship between icons of different semantic distances and their response accuracy rates, the results showed significant differences in task accuracy rates among direct, indirect, and irrelevant icons. As shown in Table 7, the average accuracy rates for the three groups of icons were 0.931, 0.769, and 0.644, respectively, with an F-value of 27.845 and a P-value far less than 0.0001, indicating statistical significance. This analysis suggests that the semantic distance of icons has an important impact on users' accuracy in identifying and using the icons (Table 8).

Table 7. ANOVA Results for Correct Rate Across Icon Semantic Distances.

Dependent Variable	Source of Variation	sample size	Mean Squares	F-Value	Significance	Welch's F-Value
Correct Rate	Direct Icon	20	0.931	0.086	F = 23.672 P = 0.000***	F = 27.845 P = 0.000***
	Indirect Icon	20	0.769	0.136		
	Irrelevant Icon	20	0.644	0.164		
	Total	60	0.781	0.176		

Table 8. NASA-TLX Scale Results.

Evaluation Dimension	Average Score for Direct Icon Task	Average Score for Indirect Icon Task	Average Score for Irrelevant Icon Task
Mental Demand	6.2	9.4	14.7
Physical Demand	3.9	5.4	9.7
Temporal Demand	9.1	10.4	14.3
Performance	12.7	12.1	8.6
Effort	7.3	9.9	14.5
Frustration	5.7	9	13.9

3.3 NASA-TLX Scale Questionnaire Data

In this study, after completing tasks interacting with icons of different semantic distances, each participant was asked to fill out a NASA-TLX questionnaire, with the specific content of the questionnaire provided in the appendix. The NASA Task Load Index (NASA-TLX) is a widely recognized tool for assessing the perceived workload during the execution of specific tasks. It allows participants to rate their perceived effort and stress levels across multiple dimensions, including mental demand, physical demand, temporal demand, performance, effort, and frustration [10 ~ 11]. The questionnaire results were compiled into a table, reflecting the average score for each dimension, providing a numerical representation of the perceived workload associated with each type of task, as detailed in Table 7. The corresponding line graph is shown in Fig. 7.

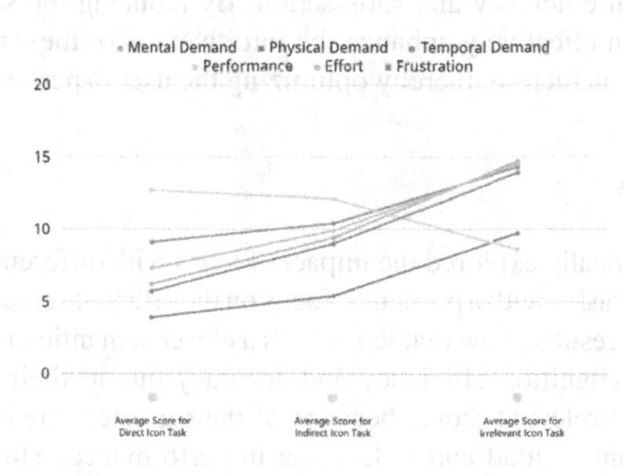

Fig. 7. Trends in NASA-TLX Scale Results.

4 Discussion

The experiment demonstrated a significant correlation between the semantic distance of icons and the participants' response time and accuracy in recognition tasks. For instance, in tasks involving icons with closer semantic distance, participants' average response time was 1637 ms, whereas this extended to 2322 ms for tasks with irrelevant icons featuring greater semantic distance, indicating an increase in cognitive load ($F = 29.674$, $P < 0.0001$). Direct icons, due to their high symbolism and clarity, enabled participants to complete recognition tasks quickly and accurately, sug-gesting that a shorter semantic distance helps reduce cognitive load. Conversely, irrel-evant icons, with their farther semantic distance, required participants to expend more cognitive resources and time, resulting in longer response times and lower accuracy rates. For example, in tasks involving irrelevant icons, the average accuracy rate was only 64.4%, significantly lower than the 93.1% for direct icons ($P < 0.0001$).

According to the NASA-TLX scale, subjective cognitive load showed clear differences across icon tasks with varying semantic distances. Participants reported higher subjec-tive cognitive load scores when dealing with irrelevant icon tasks, reflecting more com-plex and effortful cognitive processing for these types of icons. This further confirms that icons with greater semantic distance are perceived as more challenging, leading to higher subjective cognitive loads.

The conclusions of this experiment are limited by the sample size and the experimental setup. Future research should consider larger and more diverse samples, including par-ticipants from different cultural and age backgrounds. Furthermore, exploring the im-pact of different types of icon design and user interface layouts on the perception of semantic distance will aid in further refining the theory and practice of icon design. In conclusion, semantic distance is a critical factor in interface design, directly affecting user operational efficiency and satisfaction. By reducing the semantic distance of icons, designers can effectively enhance the intuitiveness of the user interface and the effi-ciency of user interaction, thereby optimizing the user experience.

5 Conclusions

This study systematically explored the impact of icons with different semantic distances on user recognition tasks, with a particular focus on the effect of semantic distance on user cognitive load. The results show that icons with a closer semantic distance signif-icantly enhanced users' recognition efficiency and accuracy due to their high intuitive-ness, while indirect and irrelevant icons, because of their greater semantic distance, led to an increase in cognitive load and a decrease in performance. This finding highlights the importance of considering semantic clarity in interface design, especially in applications that demand quick responses. Although the study involved a limited sample size, the data and analysis provided robust insights into the factors of semantic distance in icon design. These results not only help guide future interface design practices but also offer a new perspective for research in related fields.

Appendix: NASA Task Load Index (NASA-TLX) Questionnaire

NASA-TLX is an official tool used for the objective assessment of task workload. The scale takes into account multiple dimensions, including psychological, physiological, and cognitive indicators. Please fill out the questionnaire based on your actual experience in the task and accurately reflect your perceptions.

Mental Demand How mentally demanding was the task?

Very Low (0) Very High (20)

Physical Demand How physically demanding was the task?

Very Low (0) Very High (20)

Temporal Demand How hurried or rushed was the pace of the task?

Very Low (0) Very High (20)

Performance How successful were you in accomplishing what you were asked to do?

Perfect (20) Failure (0)

Effort How hard did you have to work to accomplish your level of performance?

Very Low (0) Very High (20)

Frustration How insecure, discouraged, irritated, stressed, and annoyed wereyou?

Very Low (0) Very High (20)

References

1. Lee, J.D.: Driver Distraction and Inattention: Advances in Research and Countermeasures, vol. 1. CRC Press (2017)
2. Strayer, D.L., Watson, J.M., Drews, F.A.: Cognitive distraction while multitasking in the automobile. In: Psychology of Learning and Motivation, vol. 54, pp. 29–58. Elsevier (2011). https://doi.org/10.1016/B978-0-12-385527-5.00002-4
3. Sweller, J.: Cognitive load theory. In: Psychology of Learning and Motivation, vol. 55, pp. 37–76. Elsevier (2011). https://doi.org/10.1016/B978-0-12-387691-1.00002-8
4. Engström, J., Markkula, G., Victor, T., Merat, N.: Effects of cognitive load on driving performance: the cognitive control hypothesis. Human Factors: J. Human Factors Ergon. Soc. **59**(5), 734–764 (2017), https://doi.org/10.1177/0018720817690639
5. McDougall, S.J.P., De Bruijn, O., Curry, M.B.: Exploring the effects of icon characteristics on user performance: the role of icon concreteness, complexity, and distinctiveness. J. Experim. Psychol. Appli. **6**(4), 291–306 (2000). https://doi.org/10.1037/1076-898X.6.4.291
6. Zhang, Y., Shao, J.: Research progress on the impact of icon semantic distance on cognitive load. J. Electromech. Product Developm. Innovat. **35**(4), 174–177 (2022)
7. Kuicheu, N.C., Wang, N., Tchuissang, G.N. Fanzou, Siewe, F., Xu, D.: Description logic based icons semantics: An Ontology for Icons. In: 2012 IEEE 11th International Conference on Signal Processing, pp. 1260–1263 (Oct. 2012). doi: https://doi.org/10.1109/ICoSP.2012.6491805
8. Peirce, C.S.: Collected Papers of Charles Sanders Peirce. Harvard University Press (1974)
9. Ma, Y.: Real car experiment on the influence of driver distraction duration on lane deviation. J. Jilin Univ. **04**, 45 (2015). https://doi.org/10.13229/j.cnki.jdxbgxb201504011
10. Cao, A., Chintamani, K.K., Pandya, A.K., Ellis, R.D.: NASA TLX: Software for assessing subjective mental workload
11. Hart, S.G.: NASA-Task Load Index (NASA-TLX); 20 Years Later

Pimp My Lab? The Impact of Context-Specific Usability Laboratory Environments in Public Transport

Cindy Mayas(✉), Tobias Steinert, Boris Reif, Heidi Krömker, and Matthias Hirth

Technische Universität Ilmenau, Ilmenau, Germany
{cindy.mayas,tobias.steinert,boris.reif,heidi.kroemker, matthias.hirth}@tu-ilmenau.de

Abstract. In mobile contexts, the consideration of the usage situation is important for the usability and user experience of interactive systems. When developing passenger communication systems for public transportation, it is therefore particularly relevant to take context-specific situations into account. This paper shows the challenge of designing context-specific test environments for mobile areas and analyzes their influence on user engagement and user-centered test results. Therefore, a user study with 24 participants in two independent test groups is conducted in a standard usability test environment and a context-specific test environment. In addition to single items related to the test environment, the quantitative analysis primarily reveals correlations between user engagement and the user-centered results. In contrast, the qualitative analysis, identifies differences in the content of user interactions for different situations and the resulting usability problems. In this way, the study confirms the assumption that context-specific environmental elements influence test quality, in particular for emergency situations and highly situational user needs.

Keywords: Usability Laboratory · Usability Test · Test Environment · Public Transport

1 Introduction

User studies are a central component of user-centered development (UCD). For iterative evaluations of prototypes, comparison through user studies in a controlled environment has become established [7]. However, the required efforts for creating a realistic usage environment within a laboratory is highly depended on the use case. Small office settings are usually relatively easy to realize, as they exhibit inherent similarities to a regular laboratory. In contrast, outside settings or settings within specialized environments are harder to emulate in usability laboratories.

H. Krömker (Ed.): HCII 2024, LNCS 14733, pp. 183–194, 2024.
https://doi.org/10.1007/978-3-031-60480-5_11

One area of more complex settings for usability tests are public transport vehicles. UCD has become increasing relevant also in the public transport sector [6], but conducting user field tests in the realistic usage context is challenging. On the one side, there are legislative restrictions that limit the options for recordings in public spaces, including public transport vehicles or stations. On the other side, public transport vehicles have to be taken out of operation at least temporarily to install test hardware or software. Further, there might be additional regulatory hurdles to install test hardware or software into vehicles that use public roads.

To overcome these challenges, a realistic laboratory test field is currently under construction at Technische Universität Ilmenau for enabling user studies in the context of public transport and autonomous bus shuttles in particular. This laboratory test field is part of the ÖV-LeitmotiF-KI research project[1], funded by the German Federal Ministry for Digital and Transport. The aim of the laboratory is to directly use or copy context-specific elements of public transport to create a more realistic impression of the usage environment of tasks within the usability test. In addition to requisites such as devices, furniture and buttons, digital and variable elements like wall posters and video screens are installed. Especially for applications used in mobile or changing contexts, such as passenger communication systems in public transport, the laboratory design can influence the quality of the results [3,14].

The objective of the paper is therefore to show the effects of a context-specific usability laboratory environment for the evaluation of passenger communication systems. The quality of the evaluation focuses on the user engagement during the test and the user-centered results regarding the usability and user experience (UX) of the test objects. In particular, the following research questions are derived from this aim:

RQ 1 Does the test environment influence the user-centered results?
RQ 2 Does the test environment influence the user engagement?
RQ 3 Does the user engagement influence the user-centered results?

These research questions are analyzed in a usability study with 24 participants. The usability tests are conducted in a standard laboratory (SDL) set-up and a context-specific laboratory (CSL) set-up with two separated test groups.

The remainder of this paper is structured as follows. Section 2 reviews related works. Section 3 describes the methodology of the usability tests in two different test environments. The results of the user study are presented in Sect. 4 and discussed in Sect. 5. Section 6 concludes the paper.

2 Background and Related Work

For conducting usability tests, the usage context of the tests should be as close as possible to the real context to provide valid results [10]. In addition to the

[1] https://bmdv.bund.de/SharedDocs/DE/Artikel/DG/AVF-projekte/oev-leitmotif-ki.html (German), Accessed March 2024.

formulation of the test tasks, the usage context is primarily provided by the test environment. Nevertheless, formative usability evaluations that accompany iterative UCD are often conducted in the artificial environment of usability laboratories [4] due to the higher effort and required functionality [7]. However, results from field tests can show a higher quality of results, e.g. for mobile applications [10,14]. Therefore, the challenge arises to create a usage environment within a usability laboratory as close as possible to the realistic field environment [11,13]. This offers the advantages of maintaining a laboratory environment that is easier to control and compare and at the same time supporting users to empathize with the real usage situation.

Andrzejczak et al. [1] discuss that user tests could be negatively influenced by an unfavorable test environment and the presence of an observer. In a comparative study between online tests and traditional user studies, the test participants complete two different tasks independently. During the traditional test, the test participants are observed by a test administrator. There is no direct observation during the online study. The study shows that although the test location (online vs. traditional test in the usability lab) has an influence on the test time, it has no significant influence on the results of the actual user study.

Schnöll [12] describes the essential importance of the test environment for user tests, especially in the field of autonomous driving. According to Schnöll, good results are achieved above all when either a real vehicle or a mock-up with a high level of abstraction is used for user tests. The test environment should be as real as possible for the test participants. Based on an analysis of 39 studies, Schnöll extractes eight different abstraction levels of test environments. These range from simple static simulators with few interaction options to user studies in real vehicles in real road traffic. Schnöll assesses the various levels of abstraction as part of a broad-based expert evaluation. The results determine that the higher the level of abstraction, the lower the flexibility and rapid adaptability, while the realistic impression of the test participants increases significantly.

In their investigation, Duh et al. [3] evaluates the difference between laboratory tests and field studies in the course of a study. The test subjects are mobile devices that are tested both in the laboratory and in the field. The study is based on the assumption that not all problems and aspects that can occur in real-world operations can be identified in laboratory studies. Again, the laboratory environment offers significantly better opportunities for data collection. The analysis shows that significantly more usability problems were detected in the field test than in the laboratory test. Furthermore, the frequency of problem detection is higher. The authors assume that the results in the field test would be better, as the mobile devices are designed for mobile use.

These studies indicate that the test environment could be an important factor of quality for the results of formative evaluations of passenger information and communication systems in public transport.

3 Method

3.1 Study Design

Two usability tests are performed and their results are compared to assess the influence of the lab environment. The *test object* is a passenger information system for autonomous bus shuttles in public transport. The system consists of three parts, (1) a collective station information on a non-interactive screen with dynamic data supply, (2) a collective public transport information application on an interactive touch display with static data supply, and (3) an emergency communication system establishing an audio connection to the simulated control center.

The usability test included a *standard task* and an *emergency task*. Within the standard task users fulfilled their self-controlled information needs during a 10 min-trip without event situation. The emergency task raised from an accidental car in front of the shuttle bus, which therefore could not continue the trip to the destination station. Test users were asked to fulfill their communication and assistance needs after the shuttle has stopped.

3.2 Lab Design

In the *standard laboratory*, the three parts of the test objects are arranged according to a realistic test situation with a video installation of a driving situation on a 86-inch display in front. However, other elements such as tables, chairs and cabinets remain located in the test subject's visible environment with a neutral wall design. The test environment is set up by using flexible office chairs and desks, as shown in Fig. 1. The user studies were observed by a test supervisor from an observation room. Three cameras and a semi-transparent mirror are available to the test supervisor for this purpose.

In the *context-specific laboratory*, the test situation is recreated by reducing the size of the visible test environment and using mobility-specific furniture. There are no other elements in the visible environment. The arrangement of the test situation is simulated with bus seating options and foil walls imitating the interior of a bus shuttle and videos in front and back. The realistic laboratory test field was planned and set up as a test vehicle for the arrangement and alignment of static and dynamic passenger information as well as control elements for users in a cost-efficient manner. The SDL, as shown in Fig. 2, is modeled on an automated shuttle based on the specification of an EZ10[2] shuttle from EasyMile. The basic structure of the laboratory test field consists of mechanical beams, called trusses. These offer a high load-bearing capacity for equipment such as monitors, intercoms and control units. The appearance of the vehicle interior are created by printed textile banners, which are hooked into the trusses. A cavity floor was installed in the laboratory test field, which creates a step or entrance. To increase immersion in the interior, original seating systems for local public transport vehicles are installed.

[2] https://easymile.com/vehicle-solutions/ez10-passenger-shuttle, Accessed March 2024.

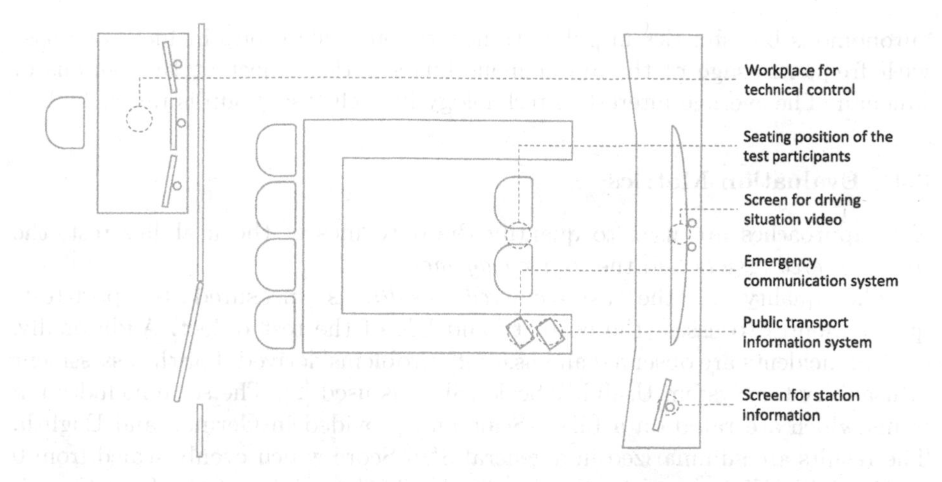

Fig. 1. Schematic representation of the standard laboratory environment

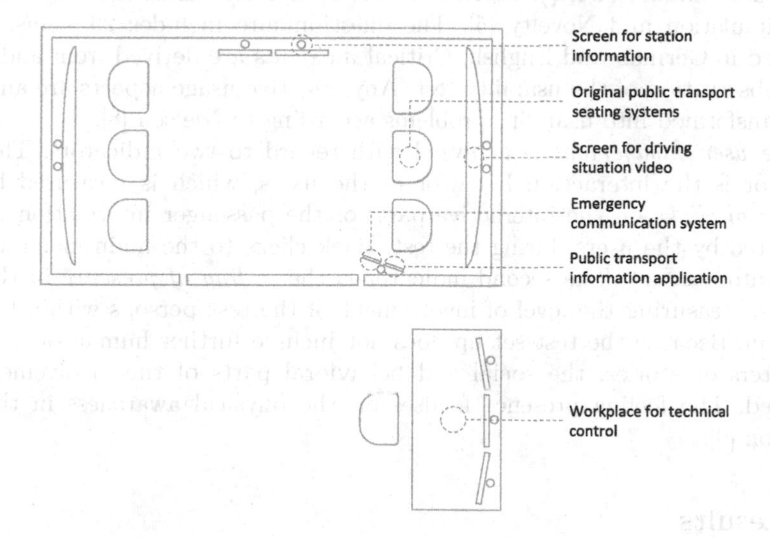

Fig. 2. Schematic representation of the context-specific laboratory environment

3.3 Test Participants

In total 24 test persons from TU Ilmenau, Germany are divided into two test groups according to the criteria of language. Each group included 6 persons conducting the usability test in English and 6 persons conducting the usability test in German. In average, test group one in the SDL environment is 26.5 years old and test group two in the CSL environment is 30.5 years old. Test group one includes 5 female and test group two 4 female persons.

Each test group includes persons with low to high experiences in using public transport. The majority of participants has no or low experiences in using autonomous bus shuttles in public transport, but each group includes one user with frequent usage of the autonomous bus shuttle project at the campus of Ilmenau. The average interest in technology in each test group is rather high.

3.4 Evaluation Metrics

Two approaches are used to quantify the outcomes of the usability test, the *user-centered results* and the *user engagement.*

The quality of the *user-centered results* is measured by post-test-questionnaires to assess the usability and UX of the test object. Additionally, critical incidents are observed and usability problems derived. For the assessment of usability the System Usability Scale (SUS) is used [2]. The scale includes ten items which are rated on a Likert-Scale and provided in German and English. The results are summarized in a general SUS-Score which evenly scaled from 0 to 100. The UX is rated by the six standardized modules of the User Experience Questionnaire (UEQ): Attractiveness, Perspicuity, Efficiency, Dependability, Stimulation and Novelty [5]. The questionnaire includes 26 items, which provided in German and English. Critical incidents are derived from audio and video observation of the usability test. Any negative usage aspects are analysed and transformed into usability problems according to Nielsen [8].

The *user engagement* is analyzed with regard to two indicators. The first indicator is the interaction behavior of the users, which is measured by the *number of clicks* on the interactive parts of the passenger information system conducted by the users during the test. Back clicks to the main menu are not taken into account. The second indicator is the *feeling of presence* of the test persons, measuring the level of involvement of the test persons within the test situation. Because the test set up does not include further human or fictitious characters or stories, the social and behavioral parts of the involvement are excluded. The feeling presence focuses on the physical awareness in the test situation [9].

4 Results

4.1 Quantitative Analysis

According to the criteria of the participants described in Sect. 3.3 the homogeneity of the test groups is confirmed with Pearson's Chi-squared test. Kolomogorov-Smirnov-Tests show normal distributions for all variables of user engagement and user-centered results. According to the small sample size, the results are checked also in a visual proof of histograms. The following Levene-Test showed a homogeneity of variances regarding the test environment groups for all variables, except the efficiency assessment in UEQ. Therefore, **RQ 1** and **RQ 2** are analyzed with a t-test for two independent samples for all dependent variables, except for the efficiency of the UEQ, which was tested with a Welch-Test for heterogenous variances.

RQ 1. The user-centered results are analyzed regarding the differences between the test groups from the SDL and the CSL in the assessed SUS-Score, the assessed UEQ modules and the number of detected usability problems per test. The majority of variables do not show significant differences ($p>.05$). The null hypothesis of no relation between the test environment and the user-centered results cannot be rejected. The UEQ efficiency is the only part of the user-centered results which shows a significant difference ($t(22) = -2.3561$; $p = .030$) between the test groups. In the SDL ($M = 1.63$; $SD = 0.58$) the efficiency of use is assessed significantly ($p<.05$) higher than in the CSL ($M = 0.88$; $SD = 0.94$). According to the following qualitative results, this might also be explained by differences in the focused contents of users' interactions.

RQ 2. The results in Table 1 show, that the test environment significantly influences parts of the user engagement. The assessment of the feeling of presence of the test persons shows significant differences in the test situations between the SDL and the CSL ($p<.01$). In contrast, the null hypothesis, that there is no difference between the test groups for the number of click interactions cannot be rejected.

Table 1. Influence of the test environment on the user engagement (**RQ 2**)

	Mean		σ		t(22)	p-value
	SDL	CSL	SDL	CSL		
Feeling presence	63.89	48.73	7.40	15.01	-3.14	.005
Number of clicks	6.92	5.67	5.90	4.14	-0.60	.554

Mean: arithmetic mean
σ: Standard Deviation (SD)
n for SDL = 12; n for CSL = 12

RQ 3. Due to the minor differences between the test groups, the analysis for this research question is conducted independently from the test environment. In the next step possible correlations are analyzed with Pearson's product-moment correlation between the user engagement of all test participants as independent variables and the quality of the user-centered results as dependent variables. While no significant correlation is detected with the number of clicks, the feeling of presence shows significant results, see Table 2. Significant positive correlation are analyzed for the feeling of presence with SUS-Score, UEQ attractiveness and stimulation ($p<.05$), as well as UEQ perspicuity and UEQ dependability ($p<.01$) and UEQ efficiency ($p<.001$).

Table 2. Correlation of user engagement and user-centered results (**RQ 3**)

	cor	t(22)	p-value	effect size
SUS-Score	0.499	2.698	.0131	medium
UEQ-Attractiveness	0.421	2.176	.0406	medium
UEQ-Perspicuity	0.575	3.297	.0032	large
UEQ-Efficiency	0.687	4.438	.0002	large
UEQ-Dependability	0.544	3.038	.006	large
UEQ-Stimulation	0.490	2.638	.015	medium
UEQ-Novelty	-0.0395	-0.182	.858	none
Usability problems	-0.071	-0.334	.7412	none

cor: r according to Pearson;
effect size: according to Cohen

The quantitative statistical analysis shows some singular significant relations between the test environment, the user engagement and the user-centered results. Nevertheless, the non-significant relations are further analyzed in a qualitative ways to show more possible dependencies, which has to be considered in the process of usability testing.

4.2 Qualitative Analysis

RQ 1. Although, the number of usability problems does not show significant differences between the two test groups, the qualitative analysis detected different issues of the problems. Table 3 shows that overall 19 usability problems are detected. 6 of theses problems are confirmed in both test environments. 7 problems were only detected from thinking aloud and observations in the SDL, further 6 problems are found in the CSL exclusively. The different usability problems in the user groups can be attributed to different user needs. The user need for information and curiosity to try out systems is higher in the SDL. In particular, more problems were discovered with regard to the interactivity and dynamic updating of information elements, such as zooming or POI functions in the city map. In contrast in the CSL, more usability problems related to the current travelling situation in the test were mentioned, such as the display of the current time and speed as well as the use of emergency communication and problem reports. This indicates a higher user need for driving comfort and safety in standard situations as well as for help and communication in emergency situation in the CSL.

Table 3. Changes in usability problems in context-specific test environment

Usability problem content	Number of different usability problems in	
	SDL	CSL
Map information	6	2
Ticket information	1	1
Legal notice	1	0
Information on station screen	4	6
Problem report	1	2
Emergency buttons	1	2
Overall	14	13

RQ 2. Considering the user engagement, the number of clicks showed no significant quantitative differences in the test groups. But the analysis of the content categories of the clicks shows, that the emergency SOS-Button is used three times in the context-specific lab and never in the standard lab. In addition, the function on anonymous problem report is used with 10 click in the CSL, in contrast to 4 clicks in the SDL, see Table 4. These interactions might indicate a higher emotional involvement in the test situation in addition to the physical presence. Especially, for information and communication systems in emergency cases a higher emotional involvement in the test situation is an important test characteristic.

Table 4. Changes in interactions in context-specific test environment

Click content category	Number of clicks in	
	SDL	CSL
Standard: Timetable	22	14
Standard: City map	29	13
Standard: Tickets	15	13
Standard: Problem report	8	11
Standard overall	77	51
Emergency: Problem report	4	10
Emergency: Info communication	5	4
Emergency: SOS communication	0	3
Emergency overall	9	17
Overall	86	68

RQ 3. The quantitative analysis shows a positive relation between the feeling of presence and the user-centered results regarding the usability assessment with SUS and the UX. However, the qualitative results of **RQ 1** and **RQ 2**

indicate similar qualitative tendencies of the results of clicks and usability problems regarding the content. Both show a higher number of results regarding the emergency situation for user interactions and usability problems. This tendency corresponds to the content-related connection between user interactions and detected usability problems in usability tests.

5 Discussion

The user study is conducted with 24 participants in two test groups. According to the research questions, the relation between the usability test environment, user engagement and user-centered results could only be shown for single items. Nevertheless, the qualitative analysis of the content of the user engagement indicated further differences, which could not be quantified. Reasons for this discrepancy are discussed in study design, the operationalization of the factors, and the test environment itself. First, the small sample size reduced the power of possible effects and represents the lower limit for the admissibility of the applied statistical tests. Second, the item battery for feeling presence is adapted from an architectural background and reduced to the applicable items. As a result, the power of the scale is reduced, which shows the Cronbach's α of 0.6 for the feeling presence scale, which was separated for this calculation in two parts of 6 items. Third, the CSL is enhanced with several typical elements from public transport, at the cost of making more technical and structural environmental factors visible for mounting and supplying these systems. The visibility of technical elements and connectors is not compared in the test environment. But this might have also influenced the feeling of the presence of the test persons, who were partly very interested in the technical background of the CSL. Hence, the feeling presence effect could not be achieved as planned. On the contrary, the context-specific elements in their first extension initially seem to create an even greater need for realistic elements, e.g. typical sounds for opening the doors.

6 Conclusion

The paper analyzes the differences between a context-specific laboratory (CSL) and standard laboratory (SDL) environment and the user engagement and user-centered results for human-computer systems in the area of public transportation, such as passenger communication systems.

Does the Test Environment Influence the User-Centered Results? (RQ1). In the two independent test groups, differences especially for the UEQ category efficiency are analyzed. The number of derived usability problems remains on an equal level, but the content of the found usability problems varies. In the CSL more usability problems according to safety needs of passengers ware revealed. In particular, the context-specific feedback on the emergency system was more detailed and heterogeneous than in the SDL. This has to be analyzed in further studies within a CSL.

Does the Test Environment Influence the User Engagement? (RQ2). Differences between the independent test groups for a SDL and CSL are revealed for feeling presence. In addition, the qualitative analysis of the content of clicks indicates a tendency to more situation and emotion related click behavior.

Does the User Engagement Influence the User-Centered Results? (RQ3). Moreover, the score for feeling presence relates to the results of SUS and five out of six UEQ categories. The relation between content of the clicks and content of the usability problems is shown qualitatively.

In the CSL, prototypes and solutions can be implemented and re-evaluated more quickly than in a field test. The CSL offers various conditions for user evaluations thanks to its flexibility and adaptability. Additional original equipment, such as handrails, straps and rescue hammers, as well as the public transport specific audio effects should be installed in the near future and evaluated in further studies.

Acknowledgments. Parts of this work were funded by the German Federal Ministry for Digital and Transport (BMDV) grant number 45AVF3004G within the project OeV-LeitmotiF-KI.

Disclosure of Interests. The authors have no competing interests to declare that are relevant to the content of this article.

References

1. Andrzejczak, C., Liu, D.: The effect of testing location on usability testing performance, participant stress levels, and subjective testing experience. J. Syst. Softw. **83**(7), 1258–1266 (2010). https://doi.org/10.1016/j.jss.2010.01.052
2. Brooke, J.: SUS: A quick and dirty usability scale. Usability Eval. Ind. **189** (11 1995)
3. Duh, H.B.L., Tan, G.C., Chen, V.H.h.: Usability evaluation for mobile device: a comparison of laboratory and field tests. In: Proceedings of the 8th Conference on Human-Computer Interaction with Mobile Devices and Services, pp. 181–186 (2006)
4. Karat, J.: Chapter 28 - User-centered software evaluation methodologies. In: Helander, M.G., Landauer, T.K., Prabhu, P.V. (eds.) Handbook of Human-Computer Interaction, 2nd edn., pp. 689–704. North-Holland, Amsterdam (1997). https://doi.org/10.1016/B978-044481862-1.50094-7, https://www.sciencedirect.com/science/article/pii/B9780444818621500947
5. Laugwitz, B., Held, T., Schrepp, M.: Construction and evaluation of a user experience questionnaire, vol. 5298, pp. 63–76 (Nov 2008). https://doi.org/10.1007/978-3-540-89350-9_6
6. Lengkong, C.R., Mayas, C., Krömker, H., Hirth, M.: The development of human-centered design in public transportation: a literature review. In: International Conference on Human-Computer Interaction. Springer (2024)

7. Mayas, C., Hörold, S., Rosenmöller, C., Krömker, H.: Evaluating methods and equipment for usability field tests in public transport. In: Kurosu, M. (ed.) HCI 2014. LNCS, vol. 8510, pp. 545–553. Springer, Cham (2014). https://doi.org/10.1007/978-3-319-07233-3_50
8. Nielsen, J.: Severity Ratings for Usability Problems: Article by Jakob Nielsen (Nov 1994). https://www.nngroup.com/articles/how-to-rate-the-severity-of-usability-problems/
9. Regenbrecht, H.: Factors for the sense of presence within virtual architecture. doctoralthesis, Bauhaus-Universität Weimar (2004). https://doi.org/10.25643/bauhaus-universitaet.33
10. Rowley, D.E.: Usability testing in the field: bringing the laboratory to the user. In: Proceedings of the SIGCHI Conference on Human Factors in Computing Systems, pp. 252–257 (1994)
11. Salzman, M.C., David Rivers, S.: Smoke and mirrors: Setting the stage for a successful usability test. Behav. Inform. Technol. **13**(1–2), 9–16 (1994)
12. Schnöll, P.: User-oriented cockpits for automated driving: a framework for the technical specification of test environments. TU Ilmenau (2022)
13. Sonderegger, A., Sauer, J.: The influence of laboratory set-up in usability tests: effects on user performance, subjective ratings and physiological measures. Ergonomics **52**(11), 1350–1361 (2009)
14. Sun, X., May, A.: A comparison of field-based and lab-based experiments to evaluate user experience of personalised mobile devices. Adv. Hum.-Comput. Interact. **2013**, 2–2 (2013)

Acceptability of Assistive Technology Promoting Independent Travel of People with Intellectual Disabilities (SAMDI PROJECT): A Focus Group Study from Support Staff

Hursula Mengue-Topio[1(✉)], Marion Duthoit[1], Laurie Letalle[1], and Youssef Guedira[2]

[1] Univ. Lille, ULR 4072 - PSITEC - Psychologie : Interactions Temps Émotions Cognition, F-59000 Lille, France
hursula.mengue-topio@univ-lille.fr

[2] LAMIH UMR CNRS 8201, Université Polytechnique Hauts-de-France, Cedex 9, 59313 Valenciennes, France

Abstract. Independent travel is essential for social participation and inclusion in the community. However, several people with intellectual and developmental disabilities restrict their community mobility to the vicinity of their residential area and face a variety of barriers to independent travel in their communities. The results of this focus group study conducted among support staff of a care organization exploring the acceptability of an assistive technology system before it is designed. According to the results, assistive technology can be a solution favored and encouraged by relatives, families, and support staff. This is however conditioned by the provision of some training on the technological tool, including for people who need greater assistance. Furthermore, the technological aid provided must be truly adapted to the cognitive and socio-emotional profile of the individuals, their needs, and their initial skill level.

Keywords: Intellectual Disabilities · Independent travel · public transportation · assistive technology · acceptability · focus group · support Staff

1 Introduction

Almost all human activities require movement inside and outside the community, thus independent travel is a prerequisite for social participation. For people with intellectual disabilities (ID) such movement may be restricted to well-known environments and these restrictions can prevent their social participation. Assistive technology systems can be a solution to improve the teaching of travel skills to persons with ID and to promote their independence and thus their social inclusion [1, 2].

ID refers to a neurodevelopmental disorder manifested by significant limitations in both intellectual functioning (IQ < 70) and adaptive behavior. Adaptive deficits are observed in conceptual, social, and practical domains. According to the Diagnostic and

H. Krömker (Ed.): HCII 2024, LNCS 14733, pp. 195–206, 2024.
https://doi.org/10.1007/978-3-031-60480-5_12

Statistical Manual of Mental Disorders (DSM)-5th edition [3] and the American Association on Intellectual and Developmental Disabilities (AAIDD) [4] this disability appears during childhood and leads to limitations in more areas, such as communication, social participation in a variety of environments such as family, professional environments and more. According to AAIDD, intellectual functioning refers to abilities such as reasoning, planning, abstract thinking, understanding complex ideas, and problem-solving. Individuals with ID face difficulties in the ability to meet expectations of autonomy and responsibility given their age and the cultural context in which they lead their lives. The causes of ID are multi-factorial [5].

Several studies about community mobility of individuals with ID show a strong restriction of independent travel in this population, which is mainly performed in the vicinity of the home and the institutions attended [6–8]. Difficulties in planning trips and managing unforeseen events or disruptions in transportation are very often mentioned by the individuals themselves. These events include route changes, delays, or human errors (wrong directions, bus or metro line; failing to get off at the right stop, and so on). Another complex problem concerns the presence of crowds during rush hour transport and human errors. In these situations, people with ID feel anxious and very often give up venturing out of their residential area or town [6, 7, 9]. The process of helping individuals with ID to become autonomous in their travel around the community is difficult, and time-consuming [9] and human assistance is often needed (families, friends, support staff, and more). Assistive technology systems could then be a solution to improve the teaching of travel skills to persons with ID and to promote their independence and thus their social inclusion [10–15]. This involves teaching specific skills such as following the appropriate procedures for using public transport to people with ID. This procedure is comprised of defining the destination, determining the departure time, knowing where to get off the bus or the metro once the destination has been reached, and following the safety rules in the event of an accident or emergency [16].

Assistive technologies designed for navigation could be a solution to help in travel assistance for people with ID. However, work on the design and use of such solutions needs to assess, the acceptability of such tools by the people concerned (people with ID, professionals, and families) before the design of these digital devices. Through the scientific literature, many factors are now better identified as having an impact on the acceptability and acceptance of these devices by these populations. These include criteria such as the perceived "need" for assistance or help, recognition of the quality of the product or usability, availability, the cost of the technological product, intended use, intrinsic motivation or reluctance to use the technology, and so on [17–20].

In addition, individual characteristics such as users' health status/level of impairment severity, and users' familiarity with assistive technology products are also other criteria to be considered in the prospective judgment toward these tools [18]. Finally, many factors such as the proper functioning of the equipment, the training of professionals in the use of the technological product (to access the location of the person being supported), and alternative solutions in case of malfunctioning are necessary so that the proposed tool is not perceived as an additional burden that would prevent its effective use [20, 24] precedent reasons show that it is essential to consider the implications for individuals around the person with ID, especially support staff [18–23]. Similarly, it is essential that the person being cared for and the support staff or family entourage easily integrate the technological assistant into their daily routines and are satisfied with it. For these reasons, [18] suggests that it is essential to consider the point of view of the person being supported, his or her professional caregivers, and/or family environment jointly when making decisions about the use of technological aids. Through scientific literature, Regarding the use of navigation aid technologies (GPS) for vulnerable populations, very few studies specifically concern the assessment of the acceptability and/or acceptance of an assistive technology for independent travel regarding people with ID.

This descriptive study is a part of the SAMDI (Mobility Assistance System for People with intellectual disability) project, which involves developing a technological aid dedicated to spatial navigation for the target population (people with ID). SAMDI is a collaborative project between interdisciplinary university research (HCI researchers and psychology researchers), care organizations for people with ID, and entrepreneurship [25]. This part of the project used the focus group method in the project to collect prospective judgments before designing the assistive technology. This technique involves group discussions based on specific themes. The focus group technique is particularly useful when exploring and discovering little-known, little-studied subjects, and allows access to the viewpoints of those concerned by the theme [26]. However, the study of mobility for people with ID remains under-researched, despite a growing number of publications in recent years. Similarly, an approach that consists of including different partners (researchers, support staff, people with ID) before, during, and after the design of an adapted navigation aid is not very common in the scientific literature. Finally, focus groups offer the advantage of low cost in terms of data collection time and financial resources. Indeed, as [27] points out, because of the difficulties of bringing together participants belonging to an identified social group (e.g. caregivers, families, people with ID) for the sole purpose of the study, group interviews may be preferred to individual interviews. In this paper, we present the approach and results of a focus group conducted with professionals who work with people with ID daily. In this descriptive study, we explore different dimensions of the acceptability of a technology before it is designed. In the literature, the acceptability and acceptance of a technological product have been studied using two major models: Davis' TAM (Technology Acceptance Model) [28] and the UTAUT model [29], which has been revised over the years [30, 31]. These two main frameworks bring together the dimensions that need to be considered to understand usage intentions and behaviors, namely perceived usefulness, perceived ease of use, the importance of previous experience, the gender or age of future users, and social influence. Despite their great influence, we have chosen to retain, in the context of this work the dimensions traditionally evaluated to measure the acceptability of a technological product by relying on these usual theoretical frameworks and other dimensions. Indeed, it is vital to understand the factors that influence the acceptability and use of navigation aid system (GPS) technologies for vulnerable groups (elderly people with or without dementia, and people with sensory or cognitive impairments following a stroke). In this context, the majority of studies focus on the use of these technologies by the elderly and people with dementia in this work [17–24].

Thus, this study relies on the perception of professionals recruited in a care organization for people with ID, who teach these people to move around daily. The focus groups were allowed to assess their prospective judgment concerning a technological aid dedicated to navigation and adapted to people with ID respect: presentation, content, proposed aids, functionalities, intended use, and reluctances.

This paper sets out the context, the method, and the main results of the focus groups with professionals. The article will conclude with a discussion of the advances and main limitations that can be drawn from this study.

2 Method

2.1 Participants

We recruited 7 French professionals (2 men and 5 women) working in a care organization "APEI du Valenciennois" among support staff. "APEI du Valenciennois" is an association located in the North of France and brings together parents and friends of people with intellectual disability, multiple and profound disabilities, autism spectrum disorders, mental disorders, or behavioral problems. They join forces to defend their rights and, in doing so, support them throughout their lives. This organization supports social inclusion whether through education, work, housing, or leisure activities. Support staff voluntarily responded to the survey. The sample included occupational therapists, social worker, social and cultural coordinator and so on. The average length of professional practice in specialized institutions was 11 years. In addition, these professionals had an average of 10 years of experience ($M = 9.58$; $SD = 7.99$) in supporting the autonomy of people with ID. Most of them worked in urban areas, in an institution located in a rural or semi-rural area, or in an industrial zone. They accompanied people with ID aged between 20 and 60 years in medico-social establishments and services. As for the level of severity of the disorders, they mentioned that they work with a population with mild to moderate ID. As regards associated disorders, we note mainly the presence of psychological disorders, behavioral disorders, communication and language disorders, and autism spectrum disorders.

Support staff who participated in the focus group were in charge of people who presented heterogeneity in their daily autonomy: some adults were able to move independently uniquely in the vicinity of the residential area, and other people were able to travel with mass transit into and around the town. Very few adults with ID can plan and travel independently to other towns.

2.2 Material

The focus group interview took the form of an organized discussion, bringing together different participants. The session was conducted by a pair consisting of an interviewer (or facilitator, moderator) and an observer. The latter noted all the verbal, non-verbal, and relational aspects that appear during the meetings [26]. The interviewer or moderator is responsible for animating the group: considering and following the emergence of new subjects and spontaneous reflections so that the participants feel that their words are considered and that their opinion is relevant [27]. The researchers used an interview guide including a set of open-ended questions concerning the support staff's opinions about the tendency of adults with ID to use an assistive technology dedicated to daily mobility, the intrinsic motivation, intended use, and reluctance to use such an aid. Other questions were about the functions, presentation of pieces of information, content, proposed aids, and functionalities of this kind of technology. We proposed about a dozen questions.

2.3 Procedure

Before the focus groups were held, the team of researchers organized various meetings to present the research Project, meet the participants (persons with ID, support staff, families), and explain the nature, interest, and methods of implementation of the project and its impact. At this stage, the researchers gave consent documents to the professionals to participate in the focus groups. For practical reasons, focus groups were held in a quiet, easily accessible, and spacious meeting room of the care organization. At the beginning of the session, the participant's agreement to the audio recording and transcription of the exchanges has been obtained, with a guarantee that the confidentiality of the exchanges will be respected. Before starting, the rules for discussion within the group were also set out at the beginning of the session to explain to the participants how the session would be conducted. The following rules for discussion within the group are also established: Limitation of the exchanges to the questions and objectives defined in the grid interview, Balanced distribution of the speaking time between participants (everyone must feel free to express themselves); Respect for all opinions and the confidentiality of the points of view expressed; Clear and explicit speaking rules to ensure the quality of the recording (speak audibly, avoid asides and side conversations, etc.) We reminded to participants that they could stop participating at any time during the session and that they had the right to ask questions to the researchers about the theme studied. Breaks and refreshments are available and proposed during the reflection session (which took approximately 2H30) to prevent fatigue, lack of attention, etc. The interviews were recorded and entirely transcribed while respecting the anonymity of the participants [27].

3 Results

The focus group technique generates diversified material (verbatim) and thus makes it possible to collect a large body of data: use of audio recordings, written notes, non-verbal qualitative information, etc. [26]. For the analysis work in this study, we (researchers) opted for a thematic analysis allowing an inductive approach to the raw data. Thematic analysis is a qualitative data analysis technique that identifies the units of meaning through the words spoken by the participants of the focus groups. This methodological approach consists of selecting and organizing word labels (themes) that condense the essential content of the participants' remarks [26]. The analysis of the data and the results of the study do not include the use of statistical tests or numbers, but the results point to selected categories and themes that reflect the major findings from the focus groups. This methodological approach focuses on the meaning of what was said and does not aim to transform the opinions or motivations expressed using statistics. To find out the importance of each dimension of acceptability assessed, a complementary study aimed at identifying the occurrence of particular behaviors, opinions, or views can be drawn up. In this case, a questionnaire survey adapted to a larger sample is recommended. The responses obtained from open-ended questions of the interview grid (used during the focus group) were analyzed with a thematic analysis and then were recoded into mutually exclusive, comprehensive categories. In this paper, we present the results concerning the acceptability of a technological aid dedicated to independent travel around the community by people with ID. The Thematic analysis of audio and video recordings, written notes, comments on the questions during the focus group conducted with support staff and have been structured around the following main categories:

1. The functions and content of the technological aid;
2. The methods used to present information appropriately and the type of media used to present this information;
3. The usability and usefulness of such a system as perceived by professionals;
4. The intrinsic motivations that would lead people with ID to use the technological aid intended use and the reluctance that would lead professionals to encourage them to use it or not.

3.1 The Functions and Content of the Technological Aid

The first thematic axis has pointed out different themes summarized in Table 1. Proposals of support staff concerned the nature of pieces of information delivered by the travel assistance assistant. Through the professionals' speeches, it emerges that information should not be given continuously but at certain specific stages of the journey; Furthermore, critical functions of a technological aid designed for independent travel are pointed out: safety alert, emergency function for contacting a professional, family member, friend, and son on; such a function would be coupled with real-time remote assistance function.

3.2 Information and Support of the Tool

In this axis, professionals formulated proposals structured around 4 main themes:

1. *Combining visual and auditory modalities to communicate instructions*: support staff has reminded that the people they accompany will save time and be able to make progress if the information is given in voice and visual mode at the same time. Combining two sensory modalities to provide information is "practical, even for readers".
2. T*ype of information provided in visual format*: professionals stated that photos of locations departure, destinations, and elements of the environment like signs, bus stops at the departure of the journey and destination are the most important pieces of information people with ID need during their trips.
3. A*uditory modality- the importance of voice synthesis*: this modality should be favored to give clear, punctual information that people can understand throughout the journey. (e.g. you need to get off the bus at x point by displaying a photograph of a key landmark in the environment close to the user's destination).
4. U*se the user's phone to implement assistive technology*: according to professionals, the personal smartphone is the support to be preferred because it is a more familiar

Table 1. Professional responses related to the functions and content of the technological aid

Thematic axis (category)	Theme	Details, suggestions from professionals
The functions and content of the technological aid	Provide the user with information about the route	Place of departure, current location and destination to be reached (visual cues such as photos)
	Provide route information only available at critical times	The user with an ID needs such information in the event of unforeseen events in the environment, such as road closures, and alternative routes, to adapt and plan an alternative journey
	Include adapted details concerning mass transit	Simplified timetables, waiting times for means of transport, journey times, and strike alerts
	Provide the function of a safety alert	Safety alerts near pedestrian crossings, traffic lights, etc. are necessary to keep the wearer alert to traffic, the environment, etc
	Provide an emergency function for contacting a professional, family member, friend, etc./ real-time remote assistance function	During the journey, the person can inform us if she. or he is "lost", we can locate them and suggest an alternative route close to their current location It is essential
	Provide a diary function (optional function)	The navigation aid system can help the user by storing dates and times of appointments The system can include an alert function to remind the user of the departure time for an appointment.)

tool for people with ID. Many users now have a smartphone. Using their smartphone will not require any additional training, unlike using another device (tablet, connected watch, etc.). Using the phone will not require any extra vigilance on the part of the person themselves or the professionals: they use this tool daily, are autonomous in keeping it up and running, and so on.

3.3 Usefulness and Usability of the Assistive Technology

Concerning this thematic axis, verbatim could be structured around 4 main themes:

1. E*ase of use of technological assistance, depending on the initial skills of the people being supported*: according to professionals, the level of reading skills acquired by people and their use of a smartphone are determining factors in how easy it is to use the technological assistance.
2. T*he need for training*: professionals have pointed out that a learning period for future users is essential in addition, they do not rule out the need for assistance from professionals, especially for the least autonomous people.
3. *Perceived usefulness*: according to professionals, this tool is useful in terms of acquisition and progression in other skills, and areas. Some people with ID who are afraid of moving in new environments, traveling to unfamiliar destinations, or people who have difficulty asking other familiar transport users for help may be reassured if they are helped appropriately with this type of tool. On the other hand, professionals predict a lack of change in travel habits among people with ID who have good routines, despite the technological tool adapted to their profile.
4. *Usability linked to various considerations*: use of technological assistance depends on the support on which the technological assistance is installed (possession of a smartphone or rather a touch-tone telephone). The link with technologies in general can limit or on the contrary, promote the use of aid. People with ID may experience possible difficulties, particularly if the tool still requires a good reading level, or if the training associated with the use of the tool is unavailable or not suitable. Finally, social considerations also determine the usability of technological assistance: the support of the tool must not be stigmatizing for people with ID, who would risk abandoning the tool.

3.4 Intrinsic Motivations, Intention of Use, and Reluctance in the Assistive Technology

This thematic axis brought together two themes we structured as follows:

1. *Intrinsic motivation, Intention to use the tool*: professionals said they were in favor of using this technological assistance if it were designed. They also said they would encourage the people with ID they support to use such a tool. One of the reasons given is the importance of independent travel for the social inclusion of the people supported and their quality of life. The second reason pointed out by support staff was the increasing difficulties in making themselves available for learning and maintaining community mobility skills. The last reason mentioned by the support staff consists in the difficulty of "finding" autonomous people with ID they support and who get "lost" using simple oral instructions. For all those reasons the design of adapted and personalized technological aids is judged favorably by professionals who say they encourage the use of such a tool among future users.
2. *Reluctance of professionals related to safety issues*: Reluctance of professionals concerned with safety reasons, in other words, attention oriented towards the application to the detriment of the environment.

4 Discussion

This study was carried out among professionals accompanying people with ID. It has enriched our knowledge of the characteristics of independent travel in this population. Also, we obtained a great insight on the most important characteristics of an adapted assistive technology dedicated to spatial navigation in their community. Previous studies showed a restrictive use of navigation aid systems by people with ID and the professionals who accompany them [1, 2]. This was the case even assistive technology could be an appropriate solution for the planning and execution of journeys as well as in case of unforeseen events or emergencies. To explore the acceptability of a designed system, the SAMDI project interviewed the professionals who support the daily mobility of these individuals. They recognized the importance of a technological assistance system dedicated solely to navigation and adapted to the cognitive profile of the users and interviewees outlined the characteristics and functions that such a tool should fulfill. Concerning the nature and modality of pieces of information delivered by the system, the results of this focus group agree with previous scientific works relating to the spatial navigation of people with ID: these individuals rely on travel routines and visual elements of the environment to plan and carry out their trips. About adapting to the environment, they highlight the need to simplify information (signs, maps, timetables, etc.) to make it easier for people with ID to find their way around and take information when using transport and making connections. Thus, the proposals for staff support agree with our knowledge of the assessment of the acceptability of assistive technology among vulnerable populations: the usefulness and usability depend on various factors, but safety reasons and the need for training are fundamental points in these projects Finally, the results from this focus group show a close link between knowledge on spatial learning and navigation which derives from laboratory studies and field surveys. These are results that show the need for and encourage interdisciplinarity: collaborations between scientific disciplines (psychology researchers, HCI researchers) and professionals [25].

One of the limitations we can formulate here concerns the fact that these results come from one single focus group, maybe three or four focus groups made up of different professionals each time would have made it possible to collect richer or more nuanced opinions. Furthermore, we are aware that these results are the perceptions of professionals, who despite their long experience and expertise developed on this subject, may have different perceptions of the issue from people with ID themselves. Another limitation of this study is that the acceptability of the targeted population is essential. Additional studies are necessary to collect their opinions, feelings, and beliefs about the acceptability of such an assistive technology system dedicated to independent travel. Overall, this descriptive study consolidates and enriches our knowledge about the cornerstones of an adapted aid designed for this target population and the tendency of support staff if essential criteria are available since the designing of the technological aid. Several collaborations between researchers stemming from various disciplines (psychology, education, computer science) as in [10, 25], persons with ID, their families, and support staff seem to be necessary to define conceptual and practical aspects of technological solutions dedicated to navigation for the targeted population.

5 Conclusions

This paper presents a focus group study among professionals accompanying people with intellectual disabilities (ID). The survey revolves around the acceptability of an assistive technology designed for independent mobility of people with ID. The results suggest the importance of designing an adapted and personalized technological aid based on previous research about the preponderance of visual landmarks at critical points of the journey. The results also point to the need for emergency functions for contacting a professional, family member, or friend, in case of unforeseen events. Finally, professionals recommend safety alerts adapted training, and support on the tool, considering users who need more assistance. Until now, human assistance seems to be the privileged solution adopted by persons with ID as well as their family/accompanying professionals when they need to move independently daily. The use of technology is very limited but may be a co-construction approach like the one adopted in the SAMDI project with the involvement of people with ID, the support staff before, during, and after the design of the technological tool can help to promote the use of a technological solution to move efficiently and independently in the community.

Acknowledgment. The work presented in this paper is the result of a collaboration between researchers from the PSITEC Laboratory of the University of Lille and the LAMIH of the University Polytechnique Hauts-de-France within the framework of the two projects: TSADI (*Technologie de Soutien à l'Apprentissage des Déplacements Indépendants*) and SAMDI (*Système d'aide à la mobilité pour les personnes présentant une déficience intellectuelle*). We would like to thank the *Maison Européenne des Sciences de l'Homme et de la Société (MESHS) Lille Nord de France* for its financial support within the framework of the TSADI project. We would also to thank the *Région Hauts de France* for their financial support within the framework of the SAMDI.

References

1. Sandjojo, J., Gebhardt, W.A., Zedlitz, A.M.E.E., Hoekman, J., den Haan, J.A., Evers, A.W.M.: Promoting independence of people with intellectual disabilities: a focus group study perspectives from people with intellectual disabilities, legal representatives, and support staff. J. Policy Pract. Intellect. Disabil. **16**, 37–52 (2019)
2. Haveman, M., Tillmann, V., Stöppler, R., Kvas, Š, Monninger, D.: Mobility and traffic abilities. J. Policy Pract. Intellect. Disabil. **10**, 289–299 (2013)
3. Marc-Antoine, C.: Mini DSM-5® : Critères diagnostiques / American Psychiatric Association ; coordination générale de la traduction française Marc-Antoine Crocq et Julien Daniel Guelfi ; directeurs de l'équipe de la traduction française Patrice Boyer, Marc-Antoine Crocq, Julien Daniel Guelfi... [et al.]. Elsevier Masson (2016)
4. American association on intellectual and developmental disabilities (AAIDD):, https://www.aaidd.org/intellectual-disability/definition, (Accessed 15 Nov 2022)
5. Inserm, I.: national de la santé et de la recherche. Déficiences intellectuelles. EDP Sciences (2016). https://www.ipubli.inserm.fr/handle/10608/6816
6. Alauzet, A., Conte, F., Sanchez, J., Velche, D.: Les personnes en situation de handicap mental, psychique ou cognitif et l'usage des transports, p. 145. INRETS, CTNERHI. (2010). https://www.lescot.ifsttar.fr/fileadmin/redaction/1_institut/1.20_sites_integres/TS2/LESCOT/documents/Projets/Rapp-finalPOTASTome2.pdf

7. Mengue-Topio, H., Courbois, Y.: L'autonomie des déplacements chez les personnes ayant une déficience intellectuelle : Une enquête réalisée auprès de travailleurs en établissement et service d'aide par le travail. Revue francophone de la déficience intellectuelle **22**, 5–13 (2011)
8. Alauzet, A.: Mobilité et handicap : Une question de point de vue. Trans. Environ. Circ. **235**, 32–33 (2017)
9. Mengue-Topio, H., Letalle, L., Courbois, Y.: Autonomie des déplacements et déficience intellectuelle : Quels défis pour les professionnels ? Revue Alter **14**(2), 99–113 (2020)
10. Letalle, L., et al.: Ontology for mobility of people with intellectual disability: building a basis of definitions for the development of navigation aid systems. In: Krömker, H. (ed.) HCII 2020. LNCS, vol. 12212, pp. 322–334. Springer, Cham (2020). https://doi.org/10.1007/978-3-030-50523-3_23
11. Davies, D.K., Stock, S.E., Holloway, S., Wehmeyer, M.L.: Evaluating a GPS-based transportation device to support independent bus travel by people with intellectual disability. Intellect. Dev. Disabil. **48**(6), 454–463 (2010)
12. Stock, S.E., Davies, D.K., Hoelzel, L.A., Mullen; R.J.: Evaluation of a GPS-based system for supporting independent use of public transportation by adults with intellectual disability. Inclusion **1**(2), 133–144. (2013)
13. Mechling, L., O'Brien, E.: Computer-based video instruction to teach students with intellectual disabilities to use public bus transportation. Educ. Training Autism Developm. Disabil. **45**(2), 230–241 (2010)
14. Price, R., Marsh, A.J., Fisher, M.H.: Teaching young adults with intellectual and developmental disabilities community-based navigation skills to take public transportation. Behav. Anal. Pract. **11**(1), 46–50 (2017)
15. Gomez, J., Montoro, G., Torrado, J.C., Plaza, A.: An Adapted wayfinding system for pedestrians with cognitive disabilities. Mob. Inf. Syst. **2015**, e520572 (2015)
16. Dever, R.B.: Habiletés à la vie communautaire : Une taxonomie. Presses Inter universitaires (1997)
17. McCreadie, C., Tinker, A.: The acceptability of assistive technology to older people. Ageing Soc. **25**(1), 91–110 (2005)
18. Williamson, B., Aplin, T., de Jonge, D., Goyne, M.: Tracking down a solution: exploring the acceptability and value of wearable GPS devices for older persons, individuals with a disability and their support persons. Disabil. Rehabil. Assist. Technol. **12**(8), 822–831 (2017)
19. McShane, R., Skelt, L.: GPS tracking for people with dementia. Working Older People **13**(3), 34–37 (2009)
20. Kearns, W.D., Rosenberg, D., West, L., Applegarth, S.P.: Attitudes and expectations of technologies to manage wandering behavior in persons with dementia. Gerontechnology **6**, 89–101 (2007)
21. Robinson, L., et al.: Balancing rights and risks: Conflicting perspectives in the management of wandering in dementia. Health Risk Soc. **9**, 389–406 (2007)
22. Landau, R., Werner, S.: Ethical aspects of using GPS for tracking people with dementia: recommendations for practice. Int. Psychogeriatr. **24**(3), 358–366 (2012)
23. Landau, R., Werner, S., Auslander, G.K., Shoval, N., Heinik, J.: Attitudes of Family and professional care-givers towards the use of GPS for tracking patients with dementia: an exploratory study. British J. Social Work **39**(4), 670–692 (2009). http://www.jstor.org/stable/23724323
24. Pot, A.M., Willemse, B.M., Horjus, S.: A pilot study on the use of tracking technology: feasibility, acceptability, and benefits for people in early stages of dementia and their informal caregivers. Aging Ment. Health **16**(1), 127–134 (2012)

25. Guedira, Y., et al.: Démarche de Conception Centrée Utilisateur de Systèmes d'Aide numériques à la Mobilité pour Personnes avec Déficience Intellectuelle. In: Adjunct Proceedings of the 34th Conference on l'Interaction Humain-Machine, IHM 2023, Troyes, France, 3–6 April, pp. 3:1–3:6. ACM (2023)
26. Desrosiers, J., Lariviere, N.: Le groupe de discussion focalisée : Application pour recueillir des informations sur le fonctionnement au quotidien des personnes avec un trouble de personnalité limite. In: Corbière, M., Larivière, N. (eds.), Méthodes qualitatives, quantitatives et mixtes dans la recherche en sciences humaines, sciales et de la santé. pp. 257-280. Presses de l'Université du Québec (2014)
27. Duffaud, S., Liébart, S.: Comment les médecins généralistes limitent-ils leurs prescriptions ? Étude qualitative par entretiens collectifs. Sante Publique **26**(3), 323–330 (2014)
28. Davis, F.D.: Perceived usefulness, perceived ease of use, and user acceptance of information technology. MIS Quart. **13**(3), 319–340 (1989)
29. Venkatesh, V., Morris, M.-G., Davis, G.-B., Davis, F.-D.: User acceptance of information technology: toward a unified view. Manag. Inf. Syst. Q. **27**(3), 425–478 (2003)
30. Venkatesh, V., Bala, H.: Technology acceptance model 3 and a research agenda on interventions. Decis. Sci. **39**, 273–315 (2008). https://doi-org.ressources-electroniques.univ-lille.fr/, https://doi.org/10.1111/j.1540-5915.2008.00192.x
31. Venkatesh, V., Thong, J.L., Xu, X.: Consumer acceptance and use of information technology: extending the unified theory of acceptance and use of technology. MIS Q., **36**(1), 157–178 (2012)

Land Vehicle Control Using Continuous Gesture Recognition on Smartwatches

Thamer Horbylon Nascimento[1(✉)], Deborah Fernandes[2], Diego Siqueira[1], Gabriel Vieira[2], Gustavo Moreira[1], Leonardo Silva[1], and Fabrizzio Soares[2(✉)]

[1] Federal Institute Goiano – Campus Iporá, Iporá, GO, Brazil
thamer.nascimento@ifgoiano.edu.br

[2] Instituto de Informática, Universidade Federal de Goiás, Goiânia, GO, Brazil
fabrizzio@inf.ufg.br

Abstract. This work introduces a method for interaction between wearable devices and ground vehicles, utilizing continuous gesture recognition on a smartwatch to control an Arduino-based vehicle. Two interaction techniques were developed and compared: in the first, the vehicle moves while the user touches the screen; in the second, a double-tap initiates or stops the vehicle. Experiments included the application of SUS and NASA TLX questionnaires to assess usability and aspects such as mental load and satisfaction. Results indicate a significant user preference for the second technique, supported by SUS scores. Additionally, NASA TLX analysis reveals positive perceptions in dimensions of mental, physical, temporal load, performance, effort, and frustration. The distribution of preferences highlights that seven users favored technique two, two opted for technique one, and one did not express a preference. This study contributes to the development of intuitive and effective interactions between wearable devices and ground vehicles, exploring the potential of continuous gesture recognition.

Keywords: Smartwatch · Continuos gesture recognition · Land Vehicle · Arduino

1 Introduction

The advancement of technology and wearable devices brings us closer to the concept of ubiquitous computing proposed by Weiser [24]. In the field of Human-Computer Interaction (HCI), the concept of embodiment is not new. Works such as Dourish's [6] on embodied interaction and Ishii and Ullmer's [11] on tangible user interfaces have explored this perspective, incorporating technology into physical objects and everyday environments, assigning new meanings to our engagement with the world.

Smartwatches are wrist-worn wearable devices used in daily life, offering not only health monitoring features and notifications but also playing fundamental

H. Krömker (Ed.): HCII 2024, LNCS 14733, pp. 207–224, 2024.
https://doi.org/10.1007/978-3-031-60480-5_13

roles in human-computer interaction [10,14]. In this context, the use of smartwatches as wearable devices and the growing popularity of do-it-yourself (DIY) hardware systems provide a solid foundation for exploring new forms of interaction between the user and the surrounding environment.

The increasing popularity of smartwatches drives the exploration of new ways to leverage these devices in different scenarios. Within this context, the interaction between these intelligent devices and land vehicles becomes increasingly significant as technology advances. Harnessing the ascendancy of smartwatches and their potential for integration with other technologies highlights the importance of creating new methods of interaction.

The successful use of smartwatches with continuous gesture recognition for interaction with other devices, as well as for gaming [9,16,17], has inspired the implementation of a control method enabling users to operate an Arduino-based ground vehicle through gestures on a smartwatch screen. This work will detail the gesture definition processes and their corresponding actions on the vehicle, continuous gesture recognition, the establishment of an efficient communication protocol between the smartwatch and the vehicle, as well as the transmission and execution of commands.

We address the system's architecture, device interactions, and the results of experiments conducted to assess the usability and effectiveness of this ground vehicle control method. By exploring this approach, we aim to make a significant contribution to the field of human-computer interaction and the evolution of ground vehicle control systems through wearable devices.

Subsequent sections will provide a thorough and organized analysis of the development, experimentation, and results of this work. Initially, we will explore related works to position our study in the current context. We will then detail the development of integrated prototypes, presenting the steps and challenges encountered in this process. Following that, we will delve into the user experiment, describing the results and discussions. Finally, we will present the conclusions, summarizing the main insights gained throughout this work.

2 Related Works

The work by Cieplik and Kasper [5] compares the development of a new controller for a jetboard with an existing portable controller from Randinn. The new controller will be gesture-based, using a smartwatch connected via Bluetooth to the jetboard for gesture recognition. This controller relies on gyroscopic technology to minimize noise from acceleration and directional speed coming from the board during operation.

Patil et al.'s work [20] aims to use the Arduino Uno to create a Bluetooth-controlled robot for material handling. The control of this robot is facilitated through the development of a mobile application, allowing users to communicate with the robot through commands to move forward, backward, right, and left.

In the work of Tombeng et al. [23], researchers innovated in car control by using a smartwatch, enabling control of existing car features through voice commands. This system connects the smartwatch to the Arduino via Bluetooth, and after the connection is established, the system validates the commands sent by the smartwatch. The features controlled by the smartwatch include starting the car engine, opening and locking the car door, and raising or lowering the car window.

The work by Mishra and Stanislaus [13] introduces a smartwatch attached to patients in remote locations, aiding in wheelchair navigation and monitoring patients' vital signs, which are then transmitted through the IoT. This smartwatch is equipped with sensors to measure health parameters such as heart rate, blood pressure, body temperature, and step count. The primary controller of the system is an Arduino UNO microcontroller, serving as the interface between input and output modules. Another work utilizing a smartwatch with the Arduino Uno microcontroller is presented by Rawf and Abdulrahman [21], focusing on the use of smartwatches for people with disabilities. Additionally, the system incorporates a DS1302 real-time clock, SD memory card, button, voice recognition module, liquid crystal display (LCD), and sound speaker. The main objective of the system is to sound the time in the specific language when the button is pressed or voice commands are issued.

The work by Butt et al. [4] focuses on the development of a ground vehicle for security applications where human presence on board is impractical. This vehicle was created using the Arduino Uno with Bluetooth connectivity through the (HC-06) module and is controlled by the user through a mobile application connected to the Bluetooth module, receiving user commands such as moving forward, backward, right, and left. The vehicle is equipped with a real-time video camera to provide information to the user and a weapon for firing against an enemy if necessary. Gomes et al. [7] presents the development of a teleoperated ground vehicle for studies in Embedded Systems and Mobile Robotics. This work was implemented on the Arduino platform for two main reasons: first, it is an Open Source platform, and second, it offers ease of connection with peripherals through Shields. At the end of the study, there is a greater understanding of the concepts presented in class/lab by the teachers and consequently increased learning for the students.

Schirme et al. [22] presents the creation of a prototype for the development of autonomous ground vehicles. The goal is to build a functional prototype with processing, radio communication, and some sensors, allowing engineers, researchers, and developers to use this system for the creation, testing, and implementation of new algorithms and techniques for autonomous navigation. Wolf et al. [25] describe the development of an unmanned ground vehicle, controlled via Wi-Fi network and based on the Arduino Uno platform. Once the computer, control board, and IP camera are on the same Wi-Fi network, the control board receives UDP packets from an application running on the computer. These packets contain information about gear, speed, and steering wheel rotation. This information is parsed, analyzed, and used to control the vehicle's motors. In the work proposed by Tombeng et al. [23], the researchers innovated in car control using a smartwatch, whose features in cars can be controlled by the smartwatch through voice commands. This system must be connected to the Arduino smartwatch via Bluetooth, and after connecting to Arduino, it will validate the command or signal sent by the smartwatch.

Nascimento et al. [18] propose the development of a method that enables the use of a smartwatch as a controller for interactive movies, using continuous gesture recognition on smartwatches. Two prototypes were created; the first utilizes gesture recognition on the smartwatch screen, and the second prototype uses a pressure touch on the smartwatch to control the movie. The chosen film for experimentation was "Black Mirror: Bandersnatch." The results indicated that, for 90% of participants, using the smartwatch is more practical and easier to control Netflix compared to using a remote control.

Furthermore, continuous gesture recognition on smartwatches has been explored as a mechanism for interaction in games. Nascimento et al. [17] developed a method for interacting with platform games using continuous gesture recognition on smartwatches as a control method. Other works by Nascimento et al. proposed methods for interacting with games and text input in virtual environments [15,19]. The results of these studies indicate that users showed receptiveness to methods employing continuous gesture recognition on smartwatches for interacting with games.

In another study conducted by Nascimento et al. [9], smartwatches with continuous gesture recognition were employed to interact with household appliances. The study utilized the "Broadlink RM Mini 3" device to send infrared commands to appliances, establishing communication with the smartwatch through a wireless network.

As evident in the literature, smartwatches are utilized for various purposes, and continuous gesture recognition has been successfully employed and well-received by users. These studies provide insights to enhance the user experience when interacting with devices using smartwatches.

However, none of these works addressed the use of smartwatches with continuous gesture recognition to interact with prototypes on Arduino boards. Therefore, this study distinguishes itself by proposing the control of a land vehicle created on an Arduino board through smartwatches with continuous gesture recognition.

3 Integrated Prototypes

The prototype consists of two fundamental parts: the smartwatch and the Arduino vehicle. The smartwatch is equipped with a touch screen and Bluetooth connectivity to transmit real-time data to the Arduino vehicle. The Arduino vehicle has a set of sensors, motors, and actuators to receive and execute commands from the smartwatch.

Gesture capture on the smartwatch screen is continuous and real-time. This means that the system can recognize a variety of gestures performed directly on the screen surface and translate them into specific commands for the vehicle.

Given that interaction is performed through gestures, we use a set of simple and intuitive gestures composed of straight lines that represent the direction in which the vehicle should move. Figure 1 illustrates the gestures and their respective actions on the vehicle.

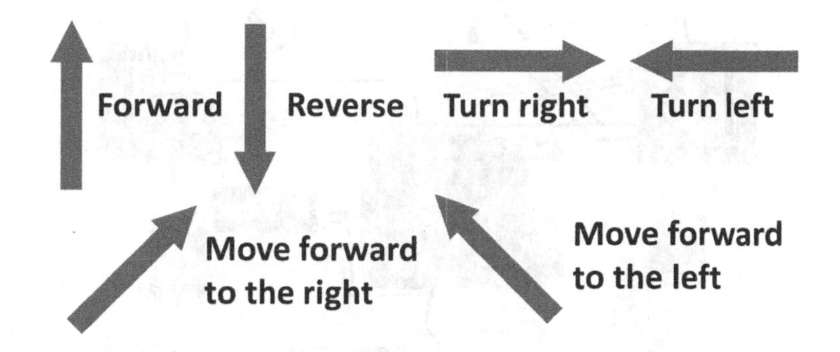

Fig. 1. Illustration of gestures and their respective actions on the vehicle.

3.1 System Architecture

With this, we have developed a method that allows the user to control a vehicle using gestures based on geometric shapes. For this, we used continuous gesture recognition on smartwatches and an Arduino-based vehicle. Figure 2 illustrates the structure of the developed method.

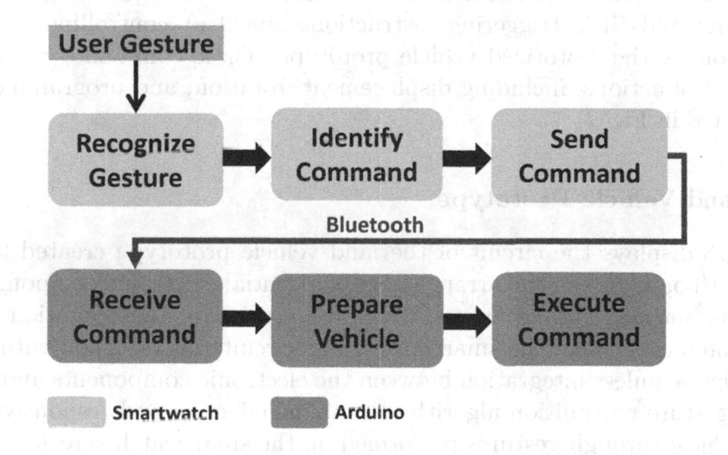

Fig. 2. Structure of the Developed Method.

As we can observe in Fig. 2, the system architecture employs an application developed for the smartwatch with the purpose of establishing a Bluetooth connection with an Arduino prototype equipped with an HC-05 module. In the context of this system, the smartwatch plays the role of continuously recognizing

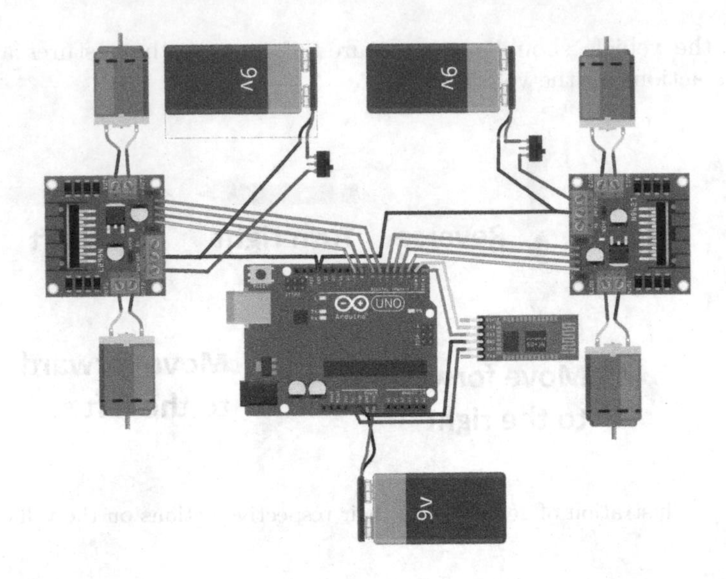

Fig. 3. Circuit Diagram of the Land Vehicle Prototype.

gestures performed by the user. Upon detecting a gesture, the smartwatch sends a text command through the Bluetooth connection to the Arduino device.

The Arduino, in turn, plays a crucial role in the system by receiving and interpreting the text command from the smartwatch via Bluetooth. It converts these commands into triggering instructions aimed at controlling movements and actions of the motorized vehicle prototype. These commands encompass a diverse set of actions, including displacement, rotation, and programmed stops. As depicted in Fig. 2.

3.2 Land Vehicle Prototype

The Fig. 3 displays the circuit of the land vehicle prototype created using the Arduino Uno, including the arrangement of essential electronic components such as motors, sensors, and actuators, which are crucial for the operation and successful interaction with the smartwatch. The circuit was designed with the aim of ensuring seamless integration between the electronic components and the continuous gesture recognition algorithm, enabling effective and responsive control of the vehicle through gestures performed on the smartwatch screen.

As we can see in Fig. 3, at the center of this system, we have the Arduino UNO board, acting as the brain of the vehicle and coordinating all operations. To enable wireless communication, we integrated a Bluetooth module HC-05, allowing remote control of the vehicle through Bluetooth connection.

The movement and direction of the vehicle are controlled by two L298N H-Bridge Driver modules, each responsible for managing two 6 V DC motors with reduction gearboxes. These motors move the wheels of the land vehicle, allowing its displacement and maneuverability. The choice of the L298N H-Bridge Driver provides precise control over the motors, ensuring efficient navigation.

The power for the system is supplied by three 9 V batteries. One of them powers the Arduino, while the other two are used to provide power to the motors, ensuring continuous operation. This power distribution is essential to avoid interference and maintain the optimal performance of the prototype.

The connections between the components are established through jumpers and switches, ensuring a solid and reliable connection. Additionally, an Arduino chassis kit is used to provide physical support to the components, ensuring stability and organization for the prototype's operation.

3.3 Smartwatch Prototype

The development of the prototype designed for WearOS-based smartwatches was a crucial step in this project. The primary functionality of the prototype involves searching for and identifying nearby Bluetooth devices. The interface displays the detected devices, allowing users to select the desired Arduino module, especially useful in scenarios where multiple Arduino vehicles are in proximity. Once the device is identified, users can establish a Bluetooth connection with the respective Arduino module.

The next step in the process involves identifying and interpreting gestures made by users. The prototype uses an advanced continuous gesture recognition algorithm. This algorithm, based on predefined models, can identify gestures in their early stages, allowing recognition even before they are fully executed. This capability optimizes the interaction between the user and the Arduino land vehicle, as the immediate detection of a gesture results in the immediate sending of a corresponding command to the connected vehicle, providing precise and agile interactions based on user gestures.

The prototype is designed to offer an intuitive and dynamic interaction between the user and the Arduino land vehicle. This process ensures not only a fast and accurate connection between the smartwatch and the vehicle but also efficient and immediate control through specific gestures recognized by the prototype.

Continuous Gesture Recognition. We utilized the continuous gesture recognition algorithm proposed by Kristensson and Denby [12] to recognize gestures performed by the user. This algorithm can recognize a gesture before it is completed with high accuracy. Considering it is possible to predict partial gestures, the user does not necessarily need to complete the entire gesture for it to be recognized; thus, it is possible to execute the desired action quickly. In our prototype, we defined that the action will be sent to the device when the user performs a gesture of at least 2 cm.

To predict partial gestures, the algorithm uses a technique that considers a gesture as a model and divides it into several segments. Thus, the segments describe partial sections of the model in increasing order [12]. Figure 4 illustrates this technique.

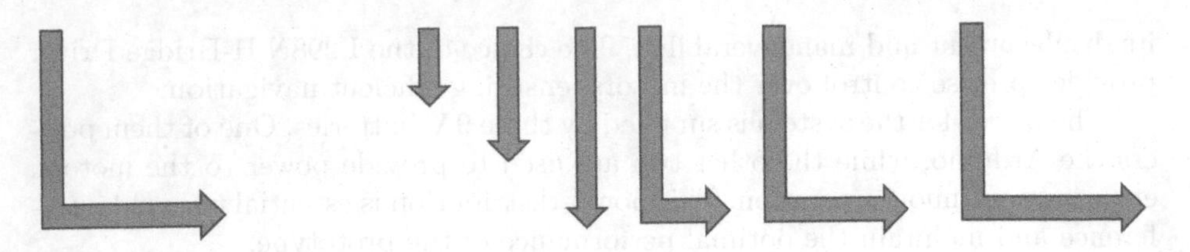

Fig. 4. Full model to the left and segments of the gesture to the right. Adapted from [12].

In Fig. 4, the direction of the gesture is represented by an arrow. A complete gesture, i.e., a model, is shown on the left with its partial segments on the right. A model can be considered as a vector of points ordered in relation to time, i.e., a vector of points ordered relative to how the movement should be produced; a gesture is segmented into several parts and increasing movements.

$$S = [s_1, s_2, ..., s_n]^T \tag{1}$$

A model is represented by w, a pair (l, S), where l is the model's description, and S is a set of segments that describes the complete model. Equation 1 describes a complete model ordered in relation to time T [12].

The continuous gesture recognition algorithm considers each gesture as a pattern to be recognized; thus, it is necessary to calculate the probability of the currently executing gesture being each gesture in the set. To reduce recognition time, we employ a multithreading gesture recognition parallelization technique proposed in our previous work [17] and successfully used in [9], to compare n gestures simultaneously, where n is the capacity of threads that the smartwatch can execute in parallel.

To enhance user interaction, we use another technique created in [17] and successfully used in [19] that identifies when the user is starting a new gesture without the need to lift their finger from the screen.

3.4 Interaction Techniques

We created and compared two interaction techniques with the vehicle, both allowing the user to control the direction of the vehicle through gestures presented in Fig. 1. The fundamental distinction between these techniques lies in the actions of starting and stopping movement. In the first technique, the vehicle moves only when the user keeps their finger on the smartwatch screen, stopping immediately when the finger is lifted. In the second technique, the user performs a double tap on the screen to start or stop the vehicle. During displacement, the vehicle follows the direction of the last received command. The crucial differences between the two techniques lie in the control methods to stop the vehicle's movement. Figure 5 illustrates the use of these techniques.

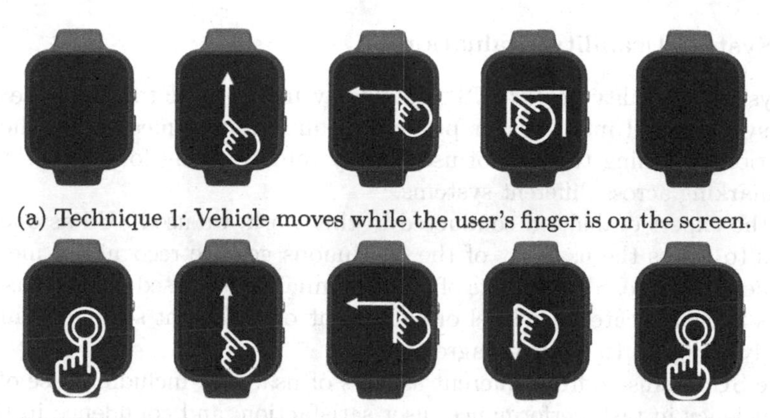

(a) Technique 1: Vehicle moves while the user's finger is on the screen.

(b) Technique 2: Double tap to start and pause the vehicle's movement.

Fig. 5. Illustration of the operation of the two developed interaction techniques on a smartwatch screen, showing the start and stop of the vehicle.

As evidenced in Subfig. 5a, in the first technique, the user keeps their finger on the smartwatch screen to control the movement of a land vehicle, with the vehicle's movement conditioned to the user's continuous touch on the screen. This approach emphasizes the feeling of continuous control, emphasizing constant tactile interaction for vehicle direction.

On the other hand, in the second technique, demonstrated in Subfig. 5b, a double tap on the smartwatch screen is used to start or stop the movement of the land vehicle. This technique stands out for its simplicity and practicality of interaction. It is important to note that both techniques use the gestures highlighted in Fig. 1 with precise gesture recognition for operation. Our experiments show the differences between these techniques and user preferences.

4 Experiment

We conducted an experiment involving 10 participants, ranging in age from 15 to 51 years. Of the participants, 8 stated that they did not own smartwatches, while 2 declared ownership. Regarding experience, 5 participants reported having no experience, 2 had little experience, 1 had regular experience, and 2 considered themselves experienced with wearable technology.

Since the smartwatch battery was not sufficient for the experiment with all users, we chose to use a smartphone, attached to the user's wrist to simulate the conditions of interaction with a smartwatch. The smartphone screen was configured to replicate the dimensions of a smartwatch, with a 1.4-inch screen. This adaptation was necessary to enable the execution of the experiment, ensuring an experience close to the reality of using a wearable device. This approach allowed for a robust evaluation of the system, considering different user profiles regarding smartwatch ownership and experience levels.

4.1 System Usability Evaluation

The System Usability Scale (SUS) is a widely used metric to assess the usability of systems and interfaces. It provides a quantitative measure of the user's perception regarding the ease of use of a system, allowing for comparison and benchmarking across different systems.

In the expert evaluation conducted in this work, we used SUS as one of the metrics to assess the usability of the continuous gesture recognition method in an immersive game. SUS consists of a questionnaire composed of 10 items, where experts must indicate their level of agreement on a 5-point scale, ranging from 'strongly disagree' to 'strongly agree.'

The SUS items address different aspects of usability, including ease of learning, efficiency in task performance, user satisfaction, and confidence in the system. The final SUS score is calculated based on the responses to the items and ranges from 0 to 100, with a higher score indicative of better perceived usability. Table 1 shows the adapted questionnaire items for our work.

Table 1. Usability questionnaire. Adapted from [2].

Q1.	I would use this system to interact with land vehicles in my daily life	↑
Q2.	I find the system unnecessarily complex	↓
Q3.	I found the system intuitive and easy to use	↑
Q4.	I think I would need the help of a person with technical knowledge to use the method	↓
Q5.	I think the various functions of the system are well integrated	↑
Q6.	I think the system exhibits a lot of inconsistency	↓
Q7.	I imagine that people will learn to use this system quickly	↑
Q8.	I found the system very complicated to use	↓
Q9.	I felt confident using the system	↑
Q10.	I needed to learn several things before using the system	↓

(↑ - Positive Question, ↓ - Negative Question).

The use of SUS in the user evaluation allows us to obtain an overall view of the usability of the continuous gesture recognition method for interacting with Arduino land vehicles. Additionally, it provides us with a comparable metric that can be used for analysis and comparison with other systems or interaction approaches.

The statements in the SUS questionnaire are alternated between positive and negative statements to avoid biased responses and allow the evaluator to analyze each question and decide whether they agree or disagree. To calculate the score from the responses, the following approach is used:

For odd-numbered responses, 1 is subtracted from the user-assigned score, i.e., $(x-1)$, where x is the user-assigned score. For even-numbered responses, the user-assigned score is subtracted from 5, i.e., $(5-x)$, where x is the user-assigned score. This approach ensures that all responses have a score in the range of 0 to 4, with 4 being the best score.

Next, the scores from all responses are multiplied by 2.5, so that each question has a score in the range of 0 to 10. The sum of the scores ranges from 0 to 100. It is important to note that this score is not a percentage but rather a general measure of usability perception. To be considered good usability, the score should be at least 68, according to Brooke [2,3].

The SUS questionnaires were administered for both interaction techniques developed in this work. Each technique was evaluated separately, resulting in a distinct SUS questionnaire for each approach. This method ensures a more specific and targeted evaluation of the usability of each technique, allowing for a more detailed comparison between the usability results of each interaction technique.

In addition to the usability evaluation using the SUS questionnaire, participants also filled out the NASA Task Load Index (NASA TLX) proposed by Hart and Staveland [8], to indicate participants' perception of mental, physical, temporal, performance, effort, and frustration loads during interaction with the control techniques. The results indicate positive user perception and provide a deeper understanding of the various elements influencing the user experience, complementing the usability analysis provided by SUS.

5 Results

In this section, we present the results of the evaluation of the two interaction techniques developed to control the Arduino ground vehicle through the continuous gesture recognition method on smartwatches. The results were obtained from the analysis of questionnaires administered to users, who evaluated each technique separately.

To assess the usability of each interaction technique, separate questionnaires were conducted for each approach. Participants were instructed to interact with the ground vehicle by applying corresponding gestures on their smartwatches as per the designated techniques. Subsequently, each participant filled out a specific questionnaire for the utilized interaction technique. The questionnaires aimed to evaluate user-friendliness, efficiency, and user satisfaction with each control method.

The results highlighted differences in usability and acceptance between the two interaction techniques. The analysis of the questionnaires revealed valuable insights into user preferences, challenges faced, and perceived effectiveness of each method. Participant scores and feedback highlighted the strengths and weaknesses of each approach, providing a deeper understanding of the performance of the techniques and guiding potential improvements.

5.1 Results of Interaction Technique 1: Active Locomotion by Touch on the Screen

We evaluated the first interaction technique, where the vehicle remains in motion while the user keeps their finger on the smartwatch screen. Figure 6 presents the average score assigned by the experts to each of the SUS questionnaire questions for this technique. The sum of the scores given by the experts reached a total of 72, indicating a positive perception of the usability of the method.

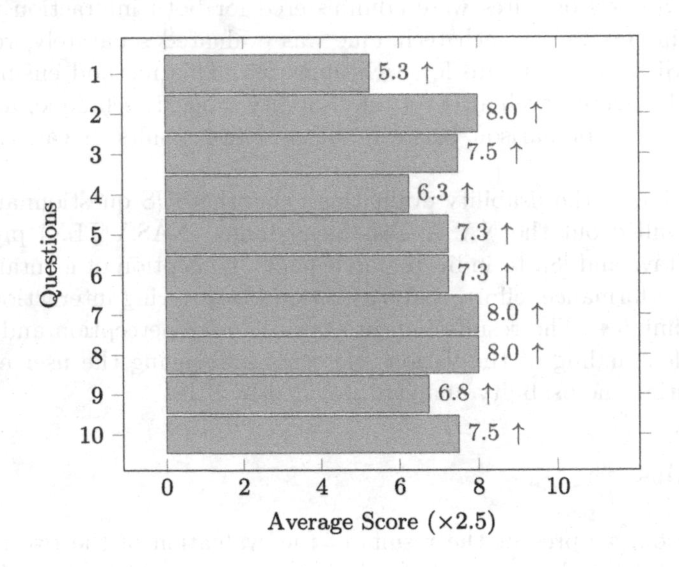

Fig. 6. Normalized average scores for each statement in the SUS questionnaire, related to Interaction Technique 1.

5.2 Results of Interaction Technique 2: Start and Stop by Double Tap

We evaluated the second interaction technique, where the user initiates and stops the vehicle's movement by double-tapping on the smartwatch screen. Figure 7 presents the average score assigned by the experts to each SUS questionnaire question for this technique. The total sum of scores reached 74.6, also indicating a positive usability perception, albeit slightly higher than that of interaction technique 1.

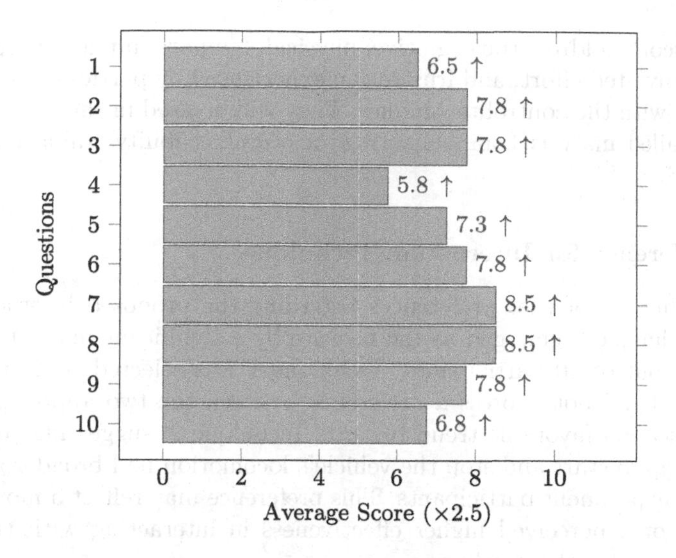

Fig. 7. Normalized average scores for each statement in the SUS questionnaire, related to Interaction Technique 2.

5.3 NASA TLX Evaluation

The results of the NASA TLX evaluation, showing participants' perceptions in different dimensions. Participants assigned scores for various aspects, and the obtained scores are as follows, as can be observed in Fig. 8.

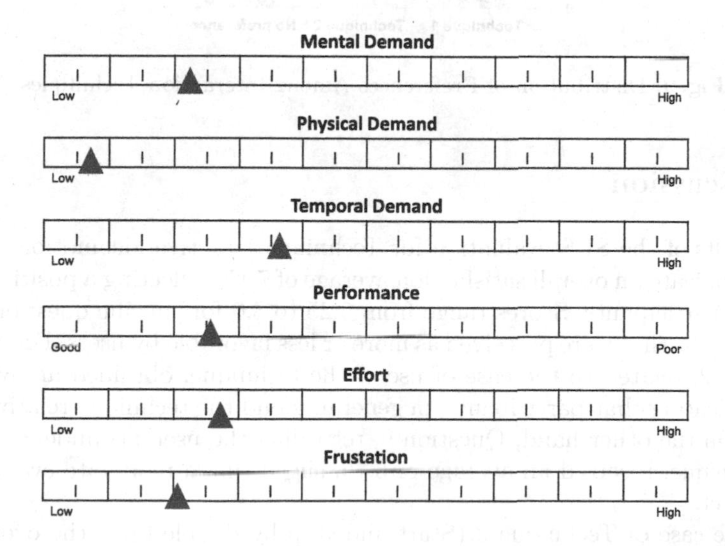

Fig. 8. Scores assigned by participants across various aspects of NASA TLX.

These scores address the cognitive, physical, temporal impact, perceived performance, invested effort, and frustration experienced by participants during the interaction with the control techniques. They will be used in the next section for a more detailed analysis, complementing the overall usability evaluation provided by the SUS.

5.4 Preference for Interaction Technique

In the evaluation of user preferences regarding the proposed interaction techniques, Technique 2 emerged as the favorite by a significant majority, with the choice of 7 out of 10 participants. Technique 1 was selected by 2 users, while 1 participant did not express a preference between the two approaches. These results indicate a favorable trend towards Technique 2, suggesting that the use of double-tap to start and stop the vehicle's locomotion had broader acceptance among the experiment participants. This preference may reflect a more intuitive experience or a perceived higher effectiveness in interacting with the vehicle. Figure 9 illustrates these values.

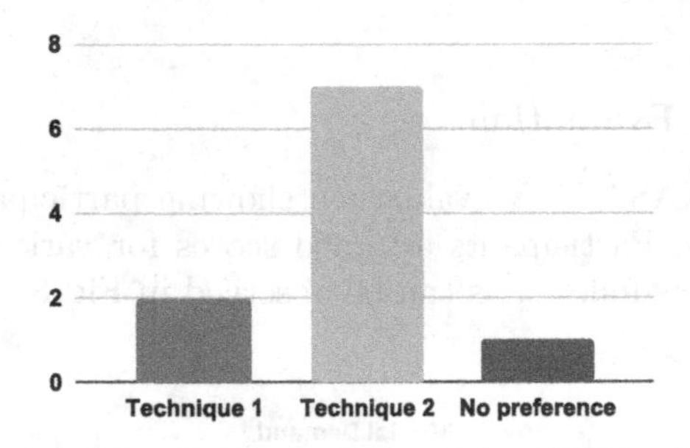

Fig. 9. Distribution of Preferences Among Interaction Techniques.

6 Discussion

The results of the SUS evaluation for Technique 1 (active locomotion by touch screen) indicate an overall satisfaction average of 7.15, reflecting a positive assessment by participants. Scores range from 5.25 to 8.0 for specific questions, highlighting areas that were perceived as more or less favorable by users. For example, Question 2, related to the ease of use of the technique, obtained an average of 7.75, indicating that participants, in general, found the technique relatively easy to use. On the other hand, Question 9, regarding the user's confidence in using the technique, received an average of 6.75, suggesting a moderate evaluation in this aspect.

In the case of Technique 2 (Start and stop by double-tap), the overall SUS average was 7.45, indicating overall positive satisfaction among participants. Scores range from 5.75 to 8.5, highlighting specific areas of strengths and areas with potential for improvement. For example, Question 7, which evaluates the ease of learning the technique, obtained an average of 8.5, indicating that users perceived the technique as easy to learn. Comparatively, Question 4, related to the need for training to use the technique, received an average of 5.75, suggesting that some participants perceived the need for more training to become fully familiar with the technique.

When comparing the two techniques, it is observed that Technique 2 received a slightly higher overall score, indicating a slight preference among participants for this approach compared to Technique 1.

The results of the experiment and usability analysis indicate that the method was well-received by users and offers a promising approach to enhance the usability and efficiency of ground vehicle control, expanding the possibilities of integrating wearable technologies with real-world devices. We hope that the method provides new experiences and contributes to the evolution of how smart devices integrate with the physical environment.

Considering the scores obtained in the SUS questionnaire, Bangor et al. [1] established a scale of descriptive adjectives based on participants' responses. This scale associates specific adjectives with the obtained scores, providing a qualitative description of the usability perception. Table 2 presents the SUS adjectives according to the score ranges. These adjectives aid in interpreting the final score and provide a general indication of the usability quality of the evaluated system.

Table 2. Descriptive Statistics of SUS Scores for Adjective Ratings. Adapted from [1].

Adjective	Mean SUS Score
Worst Imaginable	12.5
Awful	20.3
Poor	35.7
OK	50.9
Good	71.4
Excellent	85.5
Best Imaginable	90.9

The average scores for each question in the SUS questionnaires are represented in Figs. 6 and 7. The sum of the scores assigned by the experts for Technique 1 was 72, and for Technique 2, it was 74.6, indicating a positive perception of the method's usability. Although Technique 2 obtained a higher score, according to the SUS adjective scale proposed by Bangor et al. [1], both scores fall within the "Good" category.

The scores obtained by participants in various dimensions of NASA TLX detail the interaction experience with the proposed control techniques. As observed in Fig. 8, in the dimension of mental workload, the average score of 20.2 suggests a moderate level of perceived cognitive effort during the use of the techniques. This indicates that participants experienced a reasonable degree of mental demand when performing the proposed tasks, which can directly influence the perceived ease of use. The physical workload dimension has an average score of 7.2, indicating a relatively low level of physical effort associated with the interaction. This suggests that the control techniques did not impose significant physical burden on the participants.

Regarding temporal demand, the average score of 36.4 suggests that participants perceived a moderate effort related to time during task execution. This can influence the perception of method efficiency and speed. As for perceived performance, the average score of 74.6 (the left end of the chart represents ideal performance) indicates a positive evaluation by participants regarding the effectiveness of the techniques. The invested effort, with an average score of 27.2, demonstrates a moderate level of overall effort investment during interaction. Finally, the frustration dimension has an average score of 21.6, suggesting a moderate level of perceived frustration during the use of the techniques. Analyzing these scores, it is evident that the control techniques provided an overall positive experience.

6.1 Limitations

While we have developed an effective system for interaction between smartwatches and ground vehicles using continuous gesture recognition, the current scope of the prototype is specific to vehicle steering control. We have not addressed, at this stage, the implementation of additional features such as speed control, interaction with obstacles, or complex terrain variations. Expanding these functionalities would represent a significant enhancement, broadening the application possibilities of the method in different scenarios and usage contexts.

Another significant limitation is related to the variation in individual user preferences. Although we conducted experiments to assess the usability and preference of interaction techniques, preferences can be highly subjective and context-dependent. Differences in individual experiences, familiarity with wearable technologies, and personal preferences can influence participants' responses.

7 Conclusion

In this work, we presented a method of interaction between smartwatches and ground vehicles, exploring continuous gesture recognition to control the vehicle's steering. The developed prototype demonstrated effectiveness in real-time gesture capture and interpretation, providing an intuitive interaction between users and vehicles. The clear preference of participants for the technique involving double taps to initiate and stop movement highlights the importance of considering user preferences when designing interaction systems.

The results from usability assessments, using questionnaires such as SUS and NASA TLX, covered the user experience in different dimensions, including mental load, perceived effort, and overall satisfaction. Detailed analysis of NASA

TLX scores revealed nuances in participants' perceptions of specific aspects, such as temporal load and frustration, contributing to the continuous refinement of the method. The diversity of responses in SUS underscores the complexity of individual preferences, emphasizing the importance of flexibility and customization in human-technology interaction solutions. These results consolidate the relevance of the presented method, highlighting its potential and providing guidance for future research and practical improvements.

Despite the success of the method, we acknowledge limitations, including the need to expand the prototype's functionalities to encompass aspects such as speed control and interaction with obstacles. The variation in individual preferences underscores the ongoing importance of customization and adaptation of the method to meet diverse user needs.

As future work, we plan to explore these expansions and enhancements, further solidifying the viability and practical applicability of this innovative interaction method. This work represents a significant step toward more intuitive and effective integration of wearable devices with the physical world, contributing to the continuous advancement of human-computer interaction.

References

1. Bangor, A., Kortum, P., Miller, J.: Determining what individual SUS scores mean: adding an adjective rating scale. J. Usability Stud. **4**(3), 114–123 (2009). ISSN 1931-3357
2. Brooke, J.: SUS: a 'quick and dirty' usability scale. In: Usability Evaluation in Industry, 1st edn., pp. 189–194. Taylor and Francis, London (1996)
3. Brooke, J.: SUS: a retrospective. J. Usability Stud. **8**(2), 29–40 (2013). ISSN 1931-3357
4. Butt, S.U., et al.: The simple design and control of unmanned ground vehicle via Arduino and Android application: design and control of unmanned ground vehicle. Sci. Proc. Ser. **2**(1), 28–33 (2020)
5. Cieplik, J., Brinkestam, K.B.: Using hand gestures to control electric vehicles (2021)
6. Dourish, P.: Where the Action is: The Foundations of Embodied Interaction. MIT Press, Cambridge (2001)
7. Gomes, R.C., et al.: Construindo veículo teleoperado com arduino para auxilio no ensino de sistemas embarcados e robotica móvel (2012)
8. Hart, S.G., Staveland, L.E.: Development of NASA-TLX (Task Load Index): results of empirical and theoretical research. In: Hancock, P.A., Meshkati, N. (eds.) Human Mental Workload, vol. 52. Advances in Psychology, North-Holland, pp. 139–183 (1988). https://doi.org/10.1016/S0166-4115(08)62386-9, https://www.sciencedirect.com/science/article/pii/S0166411508623869
9. Horbylon Nascimento, T., Soares, F.: Home appliance control using smartwatches with continuous gesture recognition. In: Streitz, N., Konomi, S. (eds.) Distributed, Ambient and Pervasive Interactions, HCII 2021. LNCS, vol. 12782, pp. 122–134. Springer, Cham (2021). https://doi.org/10.1007/978-3-030-77015-0_9, ISBN: 978-3-030-77015-0

10. Nascimento, T.H., et al.: Interaction with smartwatches using gesture recognition: a systematic literature review. In: 2020 IEEE 44th Annual Computers, Software, and Applications Conference (COMPSAC), pp. 1661–1666 (2020). https://doi.org/10.1109/COMPSAC48688.2020.00-17
11. Ishii, H., Ullmer, B.: Tangible bits: towards seamless interfaces between people, bits and atoms. In: Proceedings of the ACM SIGCHI Conference on Human Factors in Computing Systems, CHI 1997, pp. 234–241, New York, NY, USA. Association for Computing Machinery (1997). https://doi.org/10.1145/258549.258715, ISBN 0897918029
12. Kristensson, P.O., Denby, L.C.: Continuous recognition and visualization of pen strokes and touch-screen gestures. In: EUROGRAPHICS Symposium on Sketch-Based Interfaces and Modeling (2011)
13. Mishra, A., Stanislaus, R.J.: Smart watch supported system for health care monitoring. arXiv preprint arXiv:2304.07789 (2023)
14. Moazen, D., Sajjadi, S.A., Nahapetian, A.: AirDraw: leveraging smart watch motion sensors for mobile human computer interactions. In: 2016 13th IEEE Annual Consumer Communications & Networking Conference (CCNC), pp. 442–446 (2016). https://doi.org/10.1109/CCNC.2016.7444820
15. Nascimento, T.H., et al.: Method for text input with google cardboard: an approach using smartwatches and continuous gesture recognition. In: 2017 19th Symposium on Virtual and Augmented Reality (SVR), pp. 223–226, November 2017. https://doi.org/10.1109/SVR.2017.36
16. Nascimento, T.H., et al.: Netflix control method using smartwatches and continuous gesture recognition. In: 2019 IEEE Canadian Conference of Electrical and Computer Engineering (CCECE), pp. 1–4 (2019)
17. Nascimento, T.H., et al.: Interaction with platform games using smartwatches and continuous gesture recognition: a case study. In: 2018 IEEE 42nd Annual Computer Software and Applications Conference (COMPSAC) (2018)
18. Nascimento, T.H., Soares, F.: WatchControl: a control for interactive movie using continuous gesture recognition in smartwatches. J. Inf. Process. **28**, 643–649 (2020)
19. Nascimento, T.H., et al.: MazeVR: immersion and interaction using google cardboard and continuous gesture recognition on smartwatches. In: Proceedings of the 28th International ACM Conference on 3D Web Technology, Web3D 2023, San Sebastian, Spain. Association for Computing Machinery (2023). https://doi.org/10.1145/3611314.3615912
20. Patil, V.P., et al.: Design and development of bluetooth control car based on Arduino UNO (2022)
21. Rawf, K.M.H., Abdulrahman, A.O.: Microcontroller-based Kurdish understandable and readable digital smart clock. Sci. J. Univ. Zakho **10**(1), 1–4 (2022)
22. Schirmer, A.S., et al.: Protótipo de veículo terrestre autônomo utilizando Arduino e módulo GPS. In: Revista Contemporânea **3**(6), 6377–6387 (2023)
23. Tombeng, M.T., Najoan, R., Karel, N.: Smart car: digital controlling system using android smartwatch voice recognition. In: 2018 6th International Conference on Cyber and IT Service Management (CITSM), pp. 1–5. IEEE (2018)
24. Weiser, M.: The computer for the 21st century. Sci. Am. **265**(3), 94–105 (1991). ISSN 00368733, 19467087. http://www.jstor.org/stable/24938718
25. Wolf, A.S., et al.: Veículo terrestre não tripulado controlado via rede WI-FI. In: Revista Destaques Acadêmicos, vol. 7(4) (2015)

A Study on Automotive HMI Design Evaluation Method Based on Usability Test Metrics and XGBoost Algorithm

Xiaocong Niu and Ting Tang(✉)

Ford Model e Technology (Nanjing) Co., Ltd., Shanghai Branch, Shanghai 200090, China
ttang25@ford.com

Abstract. With the development of the automotive industry, user experience research of automotive HMI design has gained increasing attention from automotive suppliers. How to discover design issues that have serious impact on user experience in the design stage is an urgent problem that need to solve. This paper proposed a new method for automotive HMI evaluation by combining usability testing data and the XGBoost algorithm. The XGBoost algorithm is utilized to construct a user complaint risk prediction model based on usability testing data. The model achieves an accuracy rate of 85.98% and an AUC value of 0.92, demonstrating good predictive performance. Feature importance analysis revealed that ease of use, cumulative driving mileage, task completion time, frequency of use, and task completion status had a greater impact on user experience, whereas path length and function type had less impact. The proposed model in this paper can be further improved in the future by combining expert evaluation methods to achieve more comprehensive and reliable label classification, and more evaluation metrics such as eye tracking data and driving behavior data can be introduced to improve the accuracy and robustness of the evaluation results.

Keywords: Automotive HMI Design Evaluation · Usability Test · XGBoost

1 Introduction

With the continuous development of automotive intelligence and connectivity, the human-machine interface (HMI) of automobiles has become increasingly complex. Automotive HMI design has become an important component of contemporary automotive design [1]. Tesla was the first to adopt a large center console touchscreen with no buttons on the Model 3, integrating navigation, music, entertainment, vehicle status, driver assistance, and other functions, subverting traditional automotive human-machine interface design [2]. In current automotive design applications, the central control screen is no longer a feature exclusive to luxury cars, and it has been extended to all models of mid-to-low-end brands. With a concurrent increase in the number and complexity of HMI elements, automotive HMI design is shifting from technology-oriented to user-oriented [3]. Automotive suppliers are increasingly focusing on the user experience of

H. Krömker (Ed.): HCII 2024, LNCS 14733, pp. 225–235, 2024.
https://doi.org/10.1007/978-3-031-60480-5_14

HMI design [4]. How to identify design issues that have a serious impact on user experience during the design phase is a pressing issue for automotive suppliers. Usability testing is one of the main methods of HMI user experience evaluation, it can get the objective results of the user's operation during the interaction process, and it is the most direct way to reflect whether the user can interact well with the product [5].

ISO 9241-110 provides some universal usability evaluation metrics, including effectiveness, efficiency, satisfaction [6], etc. In usability testing for automotive HMI design, evaluation metrics such as task completion time, task success rate, error rate, user satisfaction, complexity of operation, and user behavioral data are widely used [7]. Some studies [8] also use driving behavior data and eye-tracking data for evaluating automotive HMI design. In the process of synthesizing a comprehensive evaluation of automotive HMI design through the utilization of multiple assessment criteria, prevalent methods encompass expert evaluation, priority matrix analysis, the Analytic Hierarchy Process (AHP), and Fuzzy Comprehensive Evaluation. However, currently, there are no clear threshold specifications for these evaluation metrics, and there is no unified evaluation standard in the industry. There is a certain deviation between the results of usability test for automotive HMI design and the complaints and feedback received from actual users. Some identified usability issues may not significantly impact the user experience, and there may also be issues that users complain about which are not detected during usability testing.

This study establishes classification labels based on user complaint scores and utilizes the XGBoost algorithm for multi-metric decision-making to construct a user complaint risk prediction model based on usability testing metrics. The study explores a new method for automotive HMI evaluation design by combining usability testing metrics and the XGBoost algorithm, with the aim of promoting the development of better HMI designs that meet user needs and preferences, ultimately improving user satisfaction and reducing user complaints. In the following sections, we will introduce the usability testing method used in this study, describe the model construction method, and present our analysis results.

To mitigate the discrepancy between usability test and real user feedback, this paper introduces a new dimension of data – the user complaint rating. During usability testing, users are requested to rate the complaint level of each HMI design in the context of their own usage scenario. This study establishes classification labels based on user complaint scores and utilizes the XGBoost algorithm for multi-metric decision-making to construct a user complaint risk prediction model based on usability testing metrics. The study explores a new method for automotive HMI evaluation design by combining usability testing metrics and the XGBoost algorithm, with the aim of promoting the development of better HMI designs that meet user needs and preferences, ultimately improving user satisfaction and reducing user complaints. In the following sections, we will introduce the usability testing method used in this study, describe the model construction method, and present our analysis results.

2 Methods

2.1 Experimental Apparatus and Environment

In this experiment, we utilized a 27-inch central control screen and a Logitech G923 driving simulator including the steering wheel, pedals, and gear switching device. The visual distance—the distance from the participants' eyes to the center of the screen—ranged from 851 to 1000 mm during the experiment. The experimental prototype was developed using ProtoPie software. Figure 1 depicts the experimental environment.

Fig. 1. Experimental environment

During the experiment, a tablet computer was also utilized as a timing device. The tablet was connected to the ProtoPie prototype running on a Windows system via ProtoPie Connect software and served as a remote control for timing functions. Experimenters captured key time points of user behavior during the test by pressing the start, end, and pause buttons on the tablet interface. Video recording of the entire user testing session was conducted to facilitate subsequent analysis and review of user interactions throughout the experiment.

2.2 Participants

32 people participated in this experiment. They ranged in age from 21 to 35 years old (18 males, 14 females; average age = 28.8 years). These participants were all electric vehicle users, with driving distances ranging from 1000 to 90000 km.

2.3 Experimental Design

Experimental Tasks. This study conducted usability testing of HMI designs for commonly used functions of central control screen, covering nine functional modules such as Home, Navigation, Music, Video, Vehicle Control, System Settings, Charging Settings, Mode card, Account Settings, etc. Representative functional designs were selected for each module as user test tasks, totaling 64 tasks, and each participant performed all the tasks. Since the home page is the basis for using automobile human machine interface, the home page was taken as the first test module. Meanwhile, in order to avoid the influence of the test order on the test results, the Latin square method [9] was used to sort the other test modules. Therefore, we numbered the eight functional modules that need to be sorted as A-account settings, B-vehicle control, C-system settings, D-music, E-mode card, F-navigation, G-charging settings, and H-video. The eight modules were constructed as a Latin square matrix such that each row and column contained all the modules and no duplicate modules appeared in the same row or column. The 32 subjects are divided into 8 groups of 4 users each, and each group is assigned a test sequence based on the arrangement of the Latin square matrix.

Experimental Procedure. The testing comprised two phases: an exploratory learning phase followed by an interactive tasks testing phase, with the total duration of the test being approximately 1.5 h. Prior to commencing the test, users were briefed on the testing procedure and given detailed instructions to ensure they fully understood what was required of them.

Firstly, in the exploration and learning phase, users were permitted to freely navigate and interact with the central control screen to become acquainted with the interface and interaction design of each functional module. This phase was designed to simulate a scenario in which users are first exposed to the central control screen, and it continued until they were familiar with the operation of the corresponding functions. Upon completing the exploratory learning phase, users proceeded to the interactive task testing phase. In this phase, the experimenter issued task instructions to the users, and the users would say "start" loudly after comprehending the task content and start the operation, which was regarded as the beginning of the task. Upon believing they had finished the task, users would declare "end" to signal its completion. At these cues, the experimenter would press the "start" and "end" buttons on the tablet-based timing device to record the duration of task completion. Throughout the testing process, the experimenters documented users' operating behaviors and corresponding error paths. Following each task, users were asked to rate the HMI design based on three dimensions: frequency of use, ease of use, and degree of complaints.

2.4 Data Collection

Task Completion Status. The task completion status was divided into three categories: error-free completion, completion with errors, and failure. Completion of the task was defined as the user successfully completing the task and perceiving themselves to have completed the task. Successful completion of the task could be classified into two categories: error-free completion and completion with errors. Error-free completion was defined as the user completing the task entirely along the correct path without any errors. Completion with errors was defined as the user completing the task but making errors during the operation. If the user did not achieve the task goal or chose to give up during the testing process, it was recorded as failure.

Task Completion Time. The task completion time was recorded using a tablet-based timing tool. When the user acknowledged understanding the task by stating "start" aloud, the experimenter pressed the "start" button to log the start time of the task. Similarly, when the user indicated task completion by saying "end" out loud, the experimenter pressed the "end" button to log the end time of the task. In the event of any issues with the experiment prototype or timing tool, the experimenter would press the "break" button to flag the record for review and potential data cleansing. The duration between the recorded end time and start time of the task was calculated as the task completion time.

Subjective Score. During the testing process, after completing each test task, users evaluated the HMI designs on three dimensions: frequency of use, ease of use, and degree of complaints. These dimensions were rated on a five-point Likert scale, with the following scoring rules: for frequency of use (1-very low; 2-low; 3-moderate; 4-high; 5-very high), for ease of use (1-very poor; 2-poor; 3-average; 4-good; 5-excellent), and for the level of complaints (1-none; 2-minor; 3-moderate; 4-major; 5-severe).

3 Results and Analyses

3.1 Data Preprocessing

Data preprocessing is a crucial step in data analysis, which aims to improve the quality and reliability of the data. It mainly includes missing value processing, outlier value processing, data normalization [10], etc. In this study, missing values were primarily due to errors in recording task completion times. As these instances were few, the missing values were removed.

The boxplot method was employed to remove outliers in the task completion time data. The values that were below Q1-1.5IQR or above Q3+1.5IQR were removed from the dataset. Data normalization was performed to address the issue of performance degradation in the model caused by differences in data distribution. We compressed the data so that all feature data values were in the two-digit range. After preprocessing, the dataset included 1084 instances. The feature dataset consisted of 8 features including task completion status, task completion time, ease of use score, frequency of use score, gender, cumulative driving mileage, path length, and functional type. Gender and cumulative driving mileage are user feature data. Cumulative driving mileage refers to the total mileage accumulated by the user since the acquisition of their driver's license and can serve as an index of driving experience. The classification label was based on whether the user had complained or not, marked according to their complaint degree score. Specifically, instances with a user score of 1 were labeled as "not complained", while instances with a user score of 2, 3, 4, or 5 were labeled as "complained".

3.2 Model Construction

We utilized Extreme Gradient Boosting (XGBoost) algorithm develop a complaint prediction model. XGBoost is a scalable machine learning system for tree boosting [11], which utilizes a combination of decision trees to support parallel learning in the gradient boosting algorithm. The XGBoost algorithm combines decision trees, gradient boosting, and regularization techniques to enable fast and efficient processing of large-scale structured and sparse data. In general, the basic principle of the XGBoost algorithm is to construct multiple decision tree models and adjust the sample weights based on the error of the previous model each time a new model is trained, so that the subsequent model can fit the error better. During prediction, the results of multiple decision trees are weighted and averaged to obtain the final prediction result. Compared with the traditional GBDT algorithm, the XGBoost algorithm has the following characteristics: support for parallel computing [12], which greatly improves the training speed of the algorithm; support for missing value processing: support for L1, L2 regularization, to prevent overfitting; support for customized loss function; support for feature importance assessment. In practice, the XGBoost algorithm is widely used in feature selection, classification, regression and other fields [13], and has achieved very good results. This paper established a model based on XGBoost, with 500 decision trees, maximum depth of 3, and a learning rate of 0.0205.

3.3 Model Evaluation

K-fold Cross Validation Method. K-fold cross-validation is a widely used technique for assessing a model's generalization ability and performance. It divides the dataset into K subsets, with each subset used in turn as the validation set, while the remaining subsets are used for training. This process ensures that each subset is used for both training and testing, resulting in K models and evaluation results. In k-fold cross-validation, there is a "bias-variance trade-off" associated with the choice of k. Given these considerations, one performs k-fold cross-validation using $k = 5$ or $k = 10$ [14], as these values have been shown empirically to yield test error rate estimates that suffer neither from excessively high bias nor from very high variance. In this paper, a five-fold cross-validation method was used to divide the dataset into training and test data.

Confusion Matrix. The model can be evaluated using a confusion matrix based on the test data. The form of the confusion matrix is shown in Table 1. The confusion matrix of the prediction model is shown in Fig. 2.

Table 1. Confusion matrix

Heading level	Predicted complaint	Predicted no-complaint
True complaint	TP (true positive)	FN (false negative)
True no-complaint	FP (false positive)	TN (true negative)

Based on the confusion matrix, model evaluation metrics such as accuracy, precision, recall, and F1 score can be obtained. Accuracy refers to the proportion of samples that the model correctly classifies out of the total number of samples. It is used to assess the accuracy of the model's classifications. Precision is the ratio of true positive samples to the total number of samples predicted as positive by the model. It measures the exactness of the model in classifying a sample as positive. Recall is the proportion of true positive samples that the model correctly identifies out of the actual positive samples. It indicates

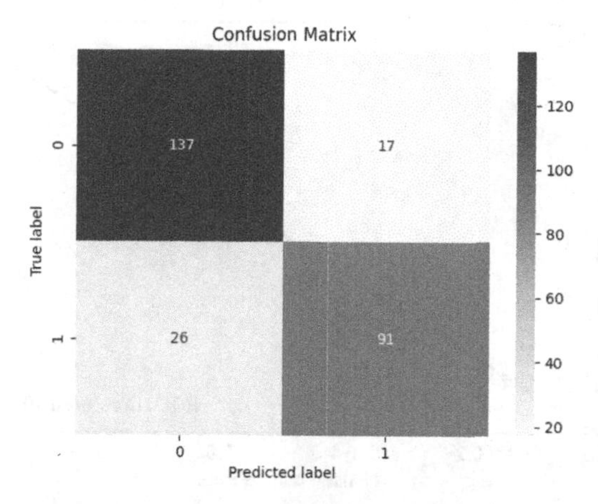

Fig. 2. Confusion matrix of the prediction model

the model's ability to identify all relevant instances. The F1 Score is a metric that combines precision and recall providing a single measure of a model's performance. It is the harmonic mean of precision and recall. They can be obtained using Eqs. (1)–(4).

$$Accuracy = \frac{TP + TN}{TP + TN + FP + FN} \tag{1}$$

$$Precision = \frac{TP}{TP + FP} \tag{2}$$

$$Recall = \frac{TP}{TP + FN} \tag{3}$$

$$F1\,Score = 2 \times \frac{Precision \times Recall}{Precision + Recall} \tag{4}$$

The results of the model evaluation metrics are shown in Table 2.

Table 2. Model Evaluation Metrics

Performance	Accuracy	Precision	Recall	F1 score
	85.98%	81.90%	84.82%	83.33%

ROC Curve. The horizontal axis of the ROC curve represents the false positive rate (FPR), which is 1-specificity, while the vertical axis represents the true positive rate (TPR), which is sensitivity. The area under the ROC curve (AUC) is a measure of a model's classification ability. A higher AUC indicates a better classification performance. The closer the ROC curve is to the upper left corner, the better the model's classification performance. The ROC curve of the model is shown in Fig. 3.

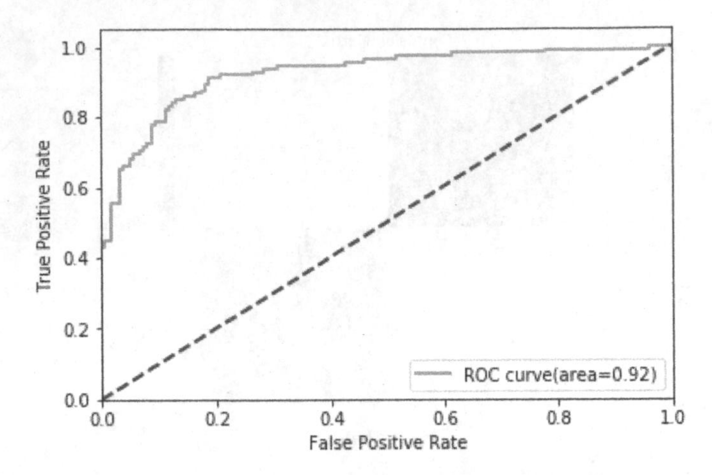

Fig. 3. ROC curve of the prediction model

The ROC curve of the complaint prediction model established in this study is close to the upper left corner, and the AUC value on the test set is 0.92, which is consistent with the cross-validation results, indicating that the model has a good classification performance.

Feature Importance Analysis. XGBoost can evaluate the contribution of each feature to the model performance by feature importance analysis. Three common methods for this analysis are based on weight, gain, and coverage. Gain represents the average information gain provided by a feature at its split points, indicating the degree of contribution of the feature. The Gain is calculated as a weighted average of the information gain at each node where the feature is used, with higher Gain values signifying a greater contribution to model performance. Weight refers to the frequency with which a feature is used across all trees, i.e., the occurrence count of the feature. A higher Weight implies that the feature is used more frequently in the model, thereby having a greater impact. Coverage denotes the number of samples affected by a feature across all trees, i.e., the feature's coverage rate. Higher Coverage indicates that the feature affects a larger number of samples, thus exerting a more substantial influence on the model. The results of the model feature importance analysis based on gain, weight, and coverage are depicted in Figs. 4, 5 and 6.

Based on the feature importance results from Cover, Gain, and Weight, it can be concluded that ease of use score, cumulative driving mileage, task completion time, frequency of use, and task completion status are important features that contribute significantly to the accuracy of the model predictions. On the other hand, gender, path length, and feature type have relatively low importance scores. Therefore, we removed these three features and retrained the model. The resulting ROC curve is shown in the Fig. 5, and the AUC value is 0.92, indicating that the removal of these features had little impact on the model results (Fig. 7).

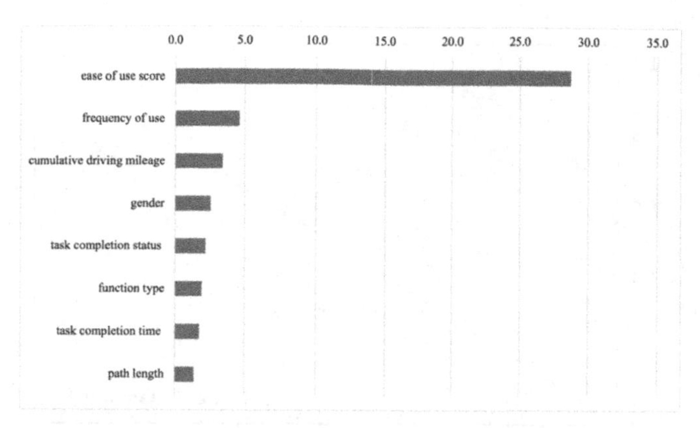

Fig. 4. Feature importance scores based on gain.

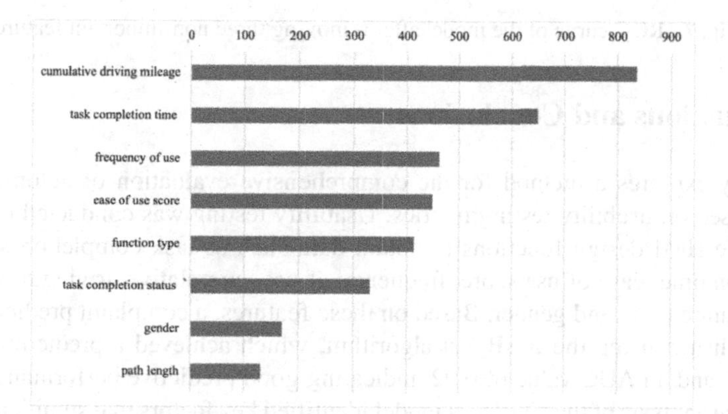

Fig. 5. Feature importance scores based on weight.

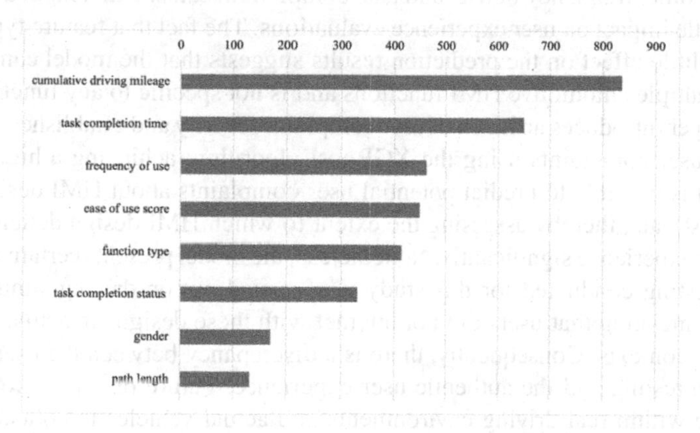

Fig. 6. Feature importance scores based on coverage.

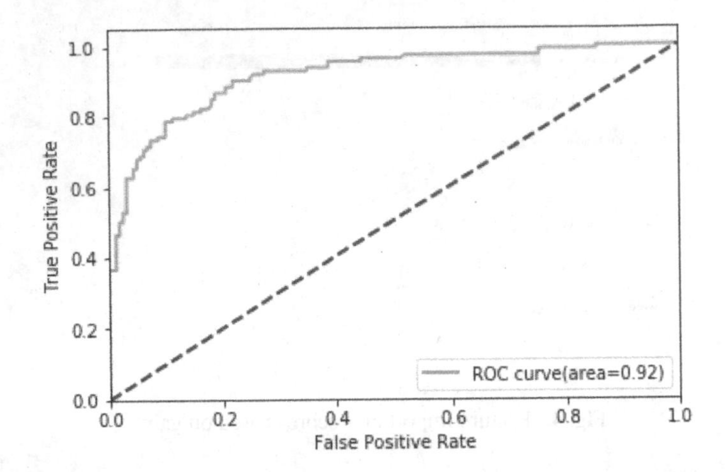

Fig. 7. ROC curve of the model after removing three non-important features

4 Discussions and Conclusions

This study explores a method for the comprehensive evaluation of automotive HMI design based on usability testing metrics. Usability testing was conducted on multiple automotive HMI design functions to obtain data such as task completion status, task completion time, ease of use score, frequency of use, cumulative driving mileage, path length, feature type, and gender. Based on these features, a complaint prediction model was established using the XGBoost algorithm, which achieved a prediction accuracy of 85.98% and an AUC value of 0.92, indicating good predictive performance. Feature importance analysis of the XGBoost model identified key factors that significantly affect the user complaints about HMI design are ease of use, cumulative driving mileage, task completion time, frequency of use, and task completion status. Path length and function type had little impact on user experience evaluations. The fact that feature type and path length had little effect on the prediction results suggests that the model can be used to evaluate multiple automotive HMI functions and is not specific to any function.

This paper introduces an index of user complaint severity and establishes a predictive model for user complaints using the XGBoost algorithm, achieving a high degree of accuracy. It is possible to predict potential user complaints about HMI design through usability test data, thereby assessing the extent to which HMI design deficiencies may impact user experience significantly. Nonetheless, the model presents certain limitations. The user testing conducted for this study was carried out on driving simulators with prototypes, meaning that users did not interact with these designs in actual vehicles or real driving contexts. Consequently, there is a discrepancy between the user complaint scores from testing and the authentic user experience. Future research should involve user testing within real driving environments and actual vehicles to collect subjective evaluations and enhance the model's applicability.

Moreover, the methods for classifying user complaint labels in this paper could be improved by incorporating expert assessments and other approaches, providing a more comprehensive and reliable system of label classification and facilitating multi-level categorization. The feature data set of the model also requires expansion. Future studies can include eye-tracking metrics and driving behavior indicators to broaden the feature data set, thereby improving the objectivity and robustness of the model.

References

1. Zhang, X., Liao, X.-P., Tu, J.-C.: A study of bibliometric trends in automotive human-machine interfaces. Sustainability. **14**, 9262 (2022). https://doi.org/10.3390/su14159262
2. Gao, H., Li, B., Han, T.: User experience research on the human-computer interaction system of connected car. Presented at the (2020)
3. Yardım, S., Pedgley, O.: Targeting a luxury driver experience: design considerations for automotive HMI and interiors (2023). https://doi.org/10.57698/V17I2.03
4. Crave, P., Francois, M., Fort, A., Osiurak, F., Jordan, N.: Automotive HMI design and participatory user involvement: review and perspectives. Ergonomics Official Publ. Ergonomics Res. Soc. **60**, 541–552 (2017)
5. Zhou, S., Jiang, L., Fan, L., Ma, J.: Build user experience evaluation system——selection and integration of indicators of automotive products. In: International Conference on Human-Computer Interaction (2022)
6. Ergonomics of Human-System Interaction Part 11: Usability: Definitions and Concepts (2018)
7. Fu, R., Zhao, X., Li, Z., Zhao, C., Wang, C.: Evaluation of the visual-manual resources required to perform calling and navigation tasks in conventional mode with a portable phone and in full-touch mode with an embedded system. Ergonomics **66**, 1633–1651 (2023). https://doi.org/10.1080/00140139.2022.2160496
8. Ma, J., Gong, Z., Tan, J., Zhang, Q., Zuo, Y.: Assessing the driving distraction effect of vehicle HMI displays using data mining techniques. Transp. Res. Part F. Traffic Psychol. Behav. **69**, 235–250 (2020)
9. Richardson, J.T.E.: The use of Latin-square designs in educational and psychological research. Educ. Res. Rev. **24**, 84–97 (2018)
10. Fu, Z., Liu, C., Peng, J., Peng, L., Qin, S.: Prediction of automobile aerodynamic drag coefficient for SUV cars based on a novel XGBoost model. Iran. J. Sci. Technol. Trans. Mech. Eng. **47**, 1349–1364 (2023)
11. Chen, T., Guestrin, C.: XGBoost: a scalable tree boosting system. In: Proceedings of the 22nd ACM SIGKDD International Conference on Knowledge Discovery and Data Mining, pp. 785–794. ACM, San Francisco, California, USA (2016)
12. Lee, E.H., Kim, K., Kho, S.Y., Kim, D.K., Cho, S.H.: Estimating express train preference of urban railway passengers based on extreme gradient boosting (XGBoost) using smart card data. Transp. Res. Rec. **2675**, 64–76 (2021)
13. Haumahu, J.P., Permana, S.D.H., Yaddarabullah, Y.: Fake news classification for Indonesian news using Extreme Gradient Boosting (XGBoost). IOP Conf. Ser. Mater. Sci. Eng. **1098**, 052081 (2021)
14. James, G., Witten, D., Hastie, T., Tibshirani, R.: An Introduction to Statistical Learning. An Introduction to Statistical Learning (2013)

Every User Has Special Needs for Inclusive Mobility

Frédéric Vanderhaegen[1,2](✉)

[1] Université Polytechnique Hauts-de-France, LAMIH UMR CNRS 8201, Le Mont Houy, 59313 Valenciennes Cedex 9, France
frederic.vanderhaegen@uphf.fr

[2] INSA Hauts-de-France, Le Mont Houy, 59313 Valenciennes Cedex 9, France

Abstract. Studies of the paper lead to confirm that each user is unique and has specific needs for inclusive mobility. They introduce a new paradigm related to inclusive Human-Computer Interaction (HCI) that aims to adapt the design of future HCI to all users whatever their cognitive, physiological, cultural, economic or cognitive levels. Three studies are presented. The first study consists in using the Reverse Comic Strip method to assess emotional impact when discovering and understanding mobility-related HCI content. It demonstrates high level of variability between people in terms of emotion and sensemaking ability. The second relates to knowledge associated to the use of advanced driving assistance system like a cruise control system and to the discovery of possible danger when using it by drivers. Here again, knowledge about system functioning and use differ between people. The last study presents a configuration of mobility of robots monitored by users who give feedback about situation awareness related to the presented HCI and their needs for future HCI. Conclusion of the paper focuses on a challenging conceptual design process of inclusive HCI to facilitate mobility for everybody and everywhere.

Keywords: Inclusive mobility · Reverse Comic Strip · Level of autonomy · Sensemaking · Situation awareness

1 Introduction

Inclusion concept usually concerns disabled, old, illiterate, disadvantaged, uncultured, troubled or vulnerable people but it can be extended to every user or worker because everybody has static and dynamic needs. From a more general viewpoint, it can concern everyone's sustainable accessibility to mobility in order to avoid any discrimination regardless of the means and the area of travel and whatever human cognitive or physical abilities [1–3]. Inclusive mobility related to freedom of movement is a fundamental human right and a sustainable lack of mobility is a source of isolation, and therefore of social exclusion. It can be studied in terms financial, physical, cognitive, educational or informational criteria.

Inclusive design principles aim to make possible employment of anyone and anywhere at each stage of the design process but also to design products that can be adapted

H. Krömker (Ed.): HCII 2024, LNCS 14733, pp. 236–248, 2024.
https://doi.org/10.1007/978-3-031-60480-5_15

to everyone and everywhere [4–8]. Their application lead to facilitate empowerment of any workers when designing products or to make easy the use of products by any users. Inclusive product and inclusive manufacturing are other similar challenges. For inclusive product design, products are implemented for all users with any experience, sensibility and abilities. Therefore, the design process considers any possible needs of users for achieving their goals. Opportunities of uses of such designed products increase and this is a high added value of the inclusive process. Inclusive manufacturing extends this concept to all stages of the product life cycle. Thus, human-system inclusion methods work on individual needs or even on needs of minorities of users while human-system integration methods mainly focus on the general interest of a majority of users [9–11].

Inclusive mobility implies the development of mobility facilities like transport networks and services, or assistive supports for indoors or outdoors application. Assistive supports are designed by applying user-centered approaches to satisfy human needs or assess human limitations [12–17]. They integrate possible different levels of automation with regard to mobility objectives and to user characteristics and constraints. Users' engagement or investment in the mobility task may therefore depend on their temporary or permanent limitations due to static or dynamic factors like fatigue, stress, workload or attention [18–21]. Such positioning of users in the control and supervisory loop of their mobility relates to Human-In-The-Loop, Human-On-the-Loop and Human-Out-Of-The-Loop concepts [22]. On the first one, users are interacting physically with the mobility mode, on the second they monitor it and intervene if necessary and on the last one, the mobility task is carried out by an entirely autonomous system, or by a third person or organization.

This paper proposes three studies related to the two first concepts and it highlights the need to take into consideration the specific needs of each user in terms of human-machine interaction to satisfy constraints and obligations of inclusive mobility. Section 2 details a study about sensemaking of a railway informational system content and analyses it by identifying variability of understanding process and associated emotional state of users. Section 3 presents variability of human skills about human and machine behaviors when driving tasks are shared between drivers and assistance systems. Section 4 explains a use-case about variability of situation awareness when humans validate intention of autonomous guided vehicles. Conclusion section makes a synthesis of results of these studies and suggests a new way to design future human-computer interactions.

2 Variability of Collective Sensemaking and Emotional State

Sensemaking consists in making relation between understanding of situations with corresponding informational supports [23, 24]. This process guides decisions or behaviors to deal with these situations. Collective or distributed sensemaking aims to converge to a common sense despite the occurrence of uncertainty levels on situations [25]. People have to rationalize jointly the understanding of a known or unknown environment by producing knowledge or rules.

This first study relates to the discovery of the meaning of data. Six groups of 2 or 5 users must make sense of the content of transport information systems that inform travelers about train or metro flows. Users were students of Master Science Quality,

Safety and Environment of Valenciennes with no specific experience on transportation domain. They have to describe their interpretation of the information system content by rules and express their emotional state during this discovery process by using the Reverse Comic Strip (RCS) method [3, 26].

The RCS method consists in defining an emotional reference associated to face pictures (see Fig. 1) and in expressing emotions by considering what is felt, said or thought. Emotional references of groups have similar interpretation for emotions like Emotion #1 or Emotion #4 but also inconsistent one like Emotion #5. Indeed, for instance, emotion associated to picture #5 is interpreted as happiness, embarrassment, or sadness. Moreover, a same picture like Emotion #5 is associated to positive or negative emotion by Group #4. These references are used for expressing users' emotional state during each exercise of sensemaking.

	Emotion #1	Emotion #2	Emotion #3	Emotion #4	Emotion #5
Group 1	angry	scared	stunned	surprised	happy
Group 2	angry	scared	doubtful	surprised	happy
Group 3	angry	shocked	incompetent	stunned	embarrassed
Group 4	angry	shocked	worried	stunned, surprised	happy, embarrassed
Group 5	irritated	scared	sad	surprised	happy
Group 6	angry	stunned	neutral, emotionless	disappointed	sad

Fig. 1. Emotional reference of each group.

Three pictures are proposed to the six groups who must discover the meaning of lit lights (see Fig. 2). After the sensemaking process about Picture #1, the real meaning is given to each group. The same is done after discovering the meaning of the content of picture #2. This manner to do is similar to the supervised learning process. For pictures #1 and #2, when lights are on, trains or metros will stop at the designated station. On picture #3, there are two subway lines, and when lights are on, there is a train stopped at the designated station.

Here again, emotions that groups expressed through their feelings, thoughts or words during each discovery exercise vary between positive, negative or neutral impressions (see Fig. 3). The discovery process for Picture #1 or for Picture #3 evokes negative emotions like scariness, anger or stun for several groups while it is mainly associated to positive emotion like happiness for Picture #2. Difficulty of sensemaking for Picture #3 is linked with expressed negative emotions and failure about the discovery process because none group discovered the correct interpretation. This can also be due to the effect of learning by conditioning because pictures have similar content.

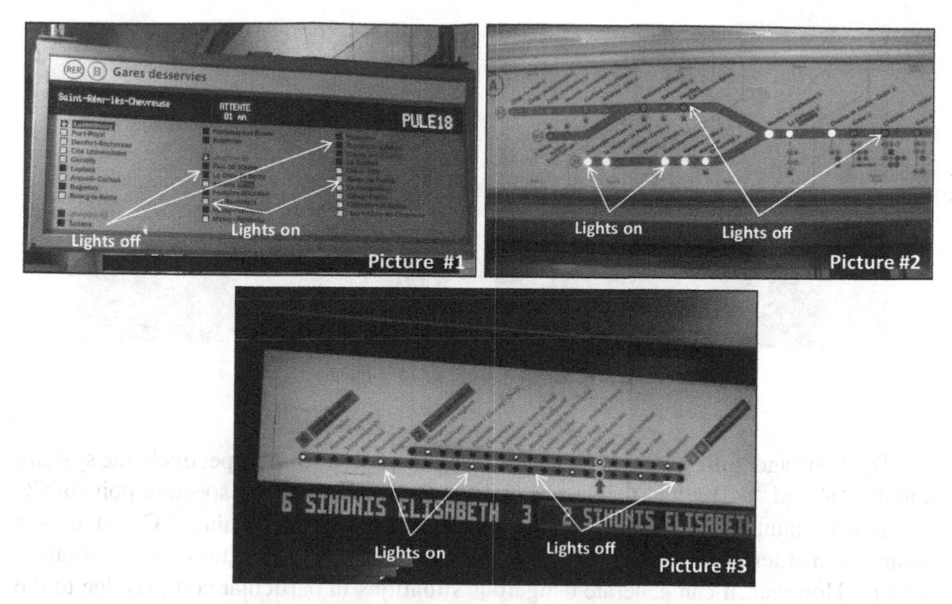

Fig. 2. Pictures #1, #2 and #3 and their content.

	Picture 1		Picture 2		Picture 3	
	Correct rule discovery?	Emotional feeling	Correct rule discovery?	Emotional feeling	Correct rule discovery?	Emotional feeling
Group 1	No	Emotions #3 + #2	No	Emotion #2 + #5	No	Emotions #2 + #3 + #1
Group 2	Yes	Emotions #4 + #3 + #2	Yes	Emotion #5	No	Emotions #4 + #3
Group 3	Yes	Emotions #4 + #3	Yes	Emotions #4 + #2	No	Emotion #1
Group 4	No	Emotion #4	Yes	Emotions #4 + #5	No	Emotions #4 + #5 + #3
Group 5	Yes	Emotions #4 + #2	No	Emotion #5	No	Emotions #1 + #3
Group 6	Yes	Emotion #3	Yes	Emotion #3	No	Emotions #2 + #3

Fig. 3. Results about interpretation of HCI content for each group.

This first study based on collective sensemaking shows that the inclusive mobility goals can fail when people cannot make correctly sense to the content of informational data system that was designed to facilitate the movement of travelers.

3 Variability of Individual Knowledge and Uncertainty

Sensemaking leads to the creation of knowledge or skills. The second study focuses on individual knowledge, the discovery of new knowledge, and the levels of certainty that subjects have about this knowledge. It is based on previous use-cases when conflicts of shared control between users and automated tools like Lane-Keeping Assist system, Cruise Control system or Adaptive Cruise Control system can occur [27–30]. It investigates about drivers' perception of the use of a Cruise Control (CC) system and of possible conflicts between their behaviors and the CC's behavior. The CC system has different buttons to make interactions with drivers possible (see Fig. 4).

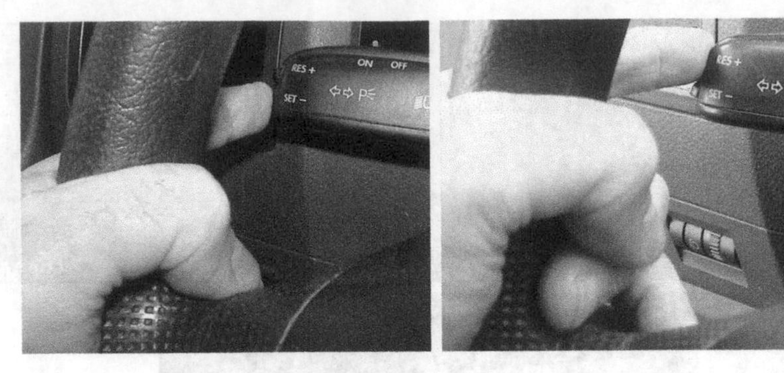

Fig. 4. HCI and use of a CC.

The "on" and "off" buttons aim at activating or deactivating respectively the system, and the "+" and "–" buttons to increase or decrease respectively the speed setpoint of CC that has to maintain the current car speed at the desired setpoint value. CC system was designed in order to avoid any over-speed that can affect car safety. It is therefore a safety system. However, it can generate dangerous situations in particular contexts due to the modification of the positioning of the legs or feet or to the use of the "+" or "–" buttons as an accelerator or braking system respectively. These effects may increase reaction time in case of emergency brake for instance. Seven drivers participated to this investigation and gave their feedback about how they behave and how they use the CC system (see Fig. 5). They are all familiar with the use and operation of CC, with aquaplaning or fuel consumption control, and had at least two years of driving experience. In order to be sure about their understanding of the CC functioning and use, the first level of feedback concerned the deactivation of the system. All drivers do not push the "on" button to deactivate the CC system but push the "off" one except for one driver. The CC system can be deactivated when pushing on the braking or clutch pedal. However, contradictions occur between the CC deactivation with the gear change and the use of the clutch pedal. Indeed, for instance upon three drivers who are sure they can deactivate the CC system by disengaging, one of them is sure that it cannot deactivate it by changing gear while this action needs to manipulate the clutch pedal. Other kinds of feedback are opposite viewpoint between drivers. This is the case of acceleration which must deactivate the CC for two drivers while four drivers disagree and one driver has a mitigated opinion. This example shows divergent and convergent feedback of drivers about their experience of the CC system use and their certainty level about this experience. Same tendencies exist for the three last groups of sentences to be validated, i.e. actions related to the modification of the speed setpoint of the CC, actions related to the safe control of an aquaplaning and action related to the fuel consumption when descending. The two last questions ask about additional functionalities and contradictions between CC use or functioning with drivers' skills. All drivers are sure or fairly sure there is not additional possible use of CC and main of them except one driver are sure there is no inconsistency between CC and driver's behaviors.

The first step of this second study demonstrates that viewpoints can converge or diverge between drivers and some of them make mistakes even if they are sure or fairly

	Absolutely yes	Fairly yes	More or less	Fairly no	Not at all	Without opinion
to disable the CC:						
I brake	6		1			
I disengage	3			2	2	
I push the "off" buttun	4	2			1	
I change gears	2	1		1	2	1
I push the "on" button					7	
I accelerate	2		1		4	
To modify speed setpoint of the CC:						
I brake		1		1	5	
I disengage		1		1	5	
I push the "+" button continuously	4		1		1	1
I push the "+" button jerkily	5	1	1			
I push the "-" button continuously	3		2	1		1
I push the "-" button jearkily	4	2	1			
To control safely an aquaplaning:						
I brake		1		2	4	
I accelerate				2	4	1
I do not brake	3	3			1	
I do not acelerate	3	2			1	1
I turn the wheel		1	1	1	3	1
I do not turn the wheel	2	3		1	1	
When going downhill, to save fuel:						
I brake			2	1	4	
I do not brake	3		2	1	1	
I take advantage of the inertia of the vehicle	5	2				
I do not take advantage of this inertia	1			4	2	
I accelerate		1	1	1	4	
I do not accelerate	3	1		2	1	
Is there any other use of CCS?				2	5	
Is there any possible inconsistency?		1			6	

Fig. 5. Feedback about driver's behaviors for driving contexts.

sure of themselves. The second step of the study proposes other sentences to be validated related to additional uses of CC and to inconsistencies between skills (see Fig. 6).

	Absolutely yes	Fairly yes	More or less	Fairly no	Not at all	Without opinion
Are you agree with these sentences:						
I can use the "+" button to accelerate	2	4		1		
I can use the "-" button to brake	2	3	1	1		
Are you agree with these inconsistencies:						
I brake to disable CC but I cannot brake in case of aquaplaning	1	2	3			1
The CC will accelerate of the current speed is under setpoint while I cannot accelerate in case of aquaplaning		2	2	1		2
In downhill, the CC will accelerate If the current speed is under the setpoint while I cannot accelerate to save fuel	1	2	1	1		2

Fig. 6. Feedback about possible new uses of CC and inconsistencies between CC and driver's behaviors.

Although opinions vary again, the majority of drivers agree with the use of the "+" or "–" buttons as an accelerator or brake system respectively. Viewpoints about the three last sentences are much more mitigated. However, some drivers are aware about possible contradictions between action on the braking pedal to deactivate CC with the fact that the activation of the braking during an aquaplaning is dangerous, or between the acceleration done by CC when current speed is under speed setpoint with the safe control of an aquaplaning or with driver's behavior to save fuel in descent.

This second study confirms that users like driver who interacts with driving assistance have their own knowledge about the situations to be managed and the HCI they use. Sometimes, the knowledge of some diverges from that of others. Such variability between users' feedback related to the production and the management of individual knowledge introduce new specific needs for controlling the correctness of sensemaking to interactions between system and human behaviors. This process will improve the development of safe inclusive mobility.

4 Variability of Situation Awareness

Situation awareness is complementary to sensemaking, i.e. the ability of production of meaning [31]. It is the ability to perceive situations, to evaluate them and to anticipate their evolution [32]. More than 70% of errors of situation awareness are due perception ability [33, 34] but a loss of situational awareness can be associated to a temporal gap between what people know before with what they know now [35, 36]. Connected watches or eye-trackers are examples of technical supports to study it [37–39].

The third study applies use-case as a support to evaluate situation awareness in terms of evaluation and anticipation of situations and of users' HCI needs. It consists in visualizing two situations of vehicle mobility (see Fig. 7). Participants to this use-case are invited to give feedback and needs about HCI. Two HCI are proposed. A control screen displays vehicle evolution. An allocation and communication interaction support displays vehicle intention to be validated or not by users (see Fig. 8). Through this support, tasks are initially allocated to vehicles that communicate to users actions they are achieving. If users do not agree with the proposed action, they select "no", otherwise they validate it by choosing "yes". They can also have no opinion on the merits or otherwise of the action by clicking on "No opinion".

On both situations, autonomous one-rail guided vehicles have limited abilities. Vehicles from Line A and B are able 1) to interpret the color of the traffic light to start at green or stop at red accordingly and 2) to avoid head-on collisions when two vehicles are less than 2 m apart. When there is a potential collision between vehicles of Line B with those of Line A, priority is given to vehicles from Line B, i.e. vehicles of Line A must stop and let those of line B pass. Vehicles from Line C have only the first ability: they stop when traffic light is red or move when it is green. Nineteen users have been invited to participate to this study.

For each operational context, they are supposed to detect possible collision between vehicles from Line A with those of Line C when traffic light of Line C is green. As a matter of fact, they might not validate the intention of V_{C2}. Several questions about the proposed HCI were asked to users.

They concern the detection of potential collision between vehicles, the situation awareness of HCI, the users' needs in terms of HCI for cooperation, competition, learning and education or pedagogy (see Fig. 9). Main users detected the collision, did not validate the V_{C2}'s intention and they thought HCI facilitated situation awareness. This

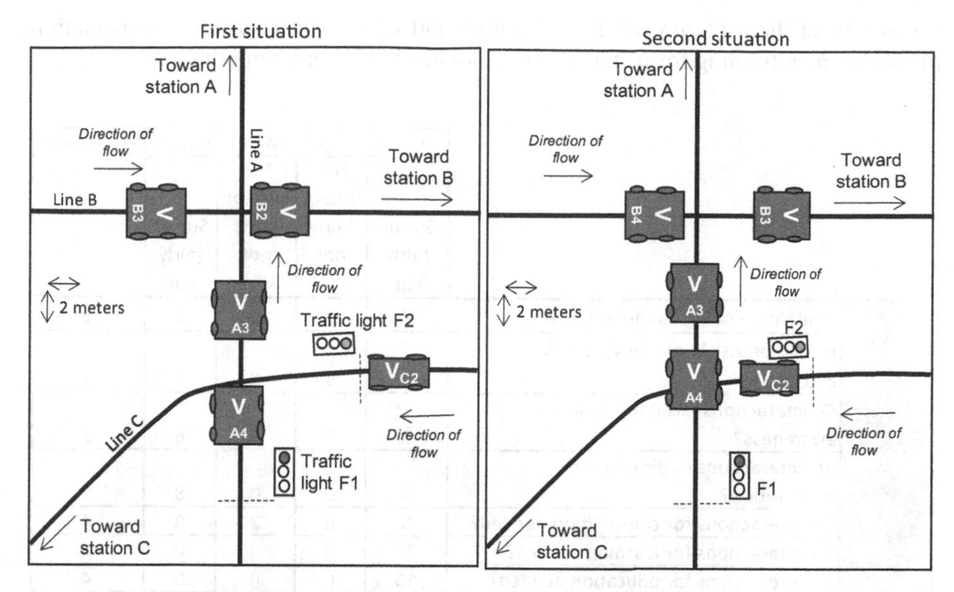

Fig. 7. Control screen content for both situations.

First situation	Destination	Vehicle intention communication	Confirmation of vehicle intention by human?
V_{A3}	Station A	Move towards the Line A / Line B intersection	Yes ☐ No ☐ No opinion ☐
V_{A4}	Station A	Pass the Line A / Line C intersection	Yes ☐ No ☐ No opinion ☐
V_{B2}	Station B	Move to the exit via station B	Yes ☐ No ☐ No opinion ☐
V_{B3}	Station B	Move towards the Line A / Line B intersection	Yes ☐ No ☐ No opinion ☐
V_{C2}	Station C	Move towards the Line A / Line C intersection	Yes ☐ No ☐ No opinion ☐

Second situation	Destination	Vehicle intention communication	Confirmation of vehicle intention by human?
V_{A3}	Station A	Stop before the crossing Line A/ Line B	Yes ☐ No ☐ No opinion ☐
V_{A4}	Station A	Stop behind R_{A3}	Yes ☐ No ☐ No opinion ☐
V_{B3}	Station B	Move to the exit via station B	Yes ☐ No ☐ No opinion ☐
V_{B4}	Station B	Move towards the Line A / Line B intersection	Yes ☐ No ☐ No opinion ☐
V_{C2}	Station C	Move towards the Line A / Line C intersection	Yes ☐ No ☐ No opinion ☐

Fig. 8. Allocation and communication interaction content for both situations.

phenomenon increased for the second situation. Regarding their feedback about HCI, they had opposite viewpoint regarding the use of proposed HCI for cooperation but a majority of them require HCI for learning and education in terms of explanation, justification, or training for instance but do not need HCI for competition.

		Yes or fairly yes		No or fairlay no		No Opinion
		Sur or fairly sur	Not sure or fairly not sure	Not sure or fairly not sure	Sur or fairly sur	
FIRST CONTEXT	Is intention of V_{C2} confirmed by user?	7	1	0	8	3
	Is collision risk between V_{C2} and V_{A4} detected?	9	3	0	6	1
	Do interactions facilitate situation awareness?	9	3	0	6	1
	Are interactions dedicated for cooperation?	9	0	0	8	2
	Are interactions for competition needed?	5	0	2	8	4
	Are interactions for learning needed?	17	0	0	0	2
	Are interactions for education needed?	15	0	0	0	4
SECOND CONTEXT	Is intention of V_{C2} confirmed by user?	4	0	0	13	2
	Is collision risk between V_{c2} and V_{A4} detected?	11	1	0	7	0
	Do interactions facilitate situation awareness?	14	1	0	3	1
	Are interactions dedicated for cooperation?	9	0	0	9	1
	Are interactions for competition needed?	5	0	0	9	5
	Are interactions for learning needed?	15	0	1	3	
	Are interactions for education needed?	13	0	0	0	6

Fig. 9. Feedback and needs for HCI.

On this third study, convergent and divergent feedback about situation awareness and HCI exist. This shows some variabilities about correct or erroneous situation awareness process and HCI needs. Technical systems to detect automatically possible collision can be developed and proposed to users, or vehicles from Line C can be equipped like vehicles of Lines A and B of a collision avoidance system. Although a majority of users required HCI for learning and education, minority points of view about competition are also relevant to consider a maximum number of needs and satisfy a maximum number of users. For instance, at each situation, five users claim with certainty the need for interactions dedicated to competition. Moreover, only nine users estimated that the proposed HCI are useful for cooperation with the system. The consideration of majority and minority viewpoint is a way to increase the inclusive mobility goal achievement, and particularly when users are active and engaged in the task of mobility.

5 Conclusion

A lot of criteria like legal obligations, preferences, needs, standard, care, accessibility are associated to the concept of inclusive mobility. This paper has proposed three different studies in order to introduced HCI new needs for inclusive mobility. It showed that although it is possible to design HCIs based on the trends or preferences of a majority of users or of groups of users, these designed HCIs will naturally be unsuitable for other minority users.

The first study presented the content of railway information systems for which user groups have to guess the meaning by applying a supervised learning process. The sense-making generated by each group may differ and erroneous meanings have sometimes occurred. This showed that the content of standard supports may not be understood by everyone and that this can constitute an obstacle to inclusive mobility. The second study concerned the evaluation of drivers' knowledge and the discovery of knowledge in the context of shared control between a driving assistance tool and drivers. Drivers who participated to the study have at least two years of driving experience and have knowledge about an aquaplaning control, fuel consumption control and cruise control system use and functioning. Despite their minimum two-year experience, knowledge about control of dynamic situations differed and some of them could be dangerous. Moreover, some drivers acknowledged the possibility to use existing HCI for achieving new goals: the use of the "+" or "−" buttons to accelerate or brake respectively instead to manage speed setpoint of the cruise control system. Due to this variability of knowledge about known or unknown situations, inclusive mobility may be impacted and new HCI has to consider this variability. The last study was about situation awareness of HCI to control flows between vehicles and about the identification of specific needs of users. Participants of this study did not have same abilities for evaluating and anticipating situations and same needs for future HCI. Majority of them required HCI for learning and education in terms of explanation, justification or training when interacting with vehicles. Some of them required the development of HCI for improving cooperation between users and vehicles and for competition in order to facilitate human engagement and creativity to the tasks. The lack of abilities of vehicles related to cooperation, competition, learning or education is another obstacle for inclusive mobility.

As a matter of fact, future HCI development has to consider variability in terms of sensemaking, knowledge and situation awareness and include specific HCI for cooperation, competition, learning and education between humans and technical smart systems.

This is a way to study human-machine co-evolution concept presented on [40] and to implement mutually human-supported computer and computer-supported human principles by making cooperation, competition, learning and education possible [10, 11]. Users are not obliged to cooperate with so-called cooperative systems like collaborative robots and some can compete with them in order to try to do better.

This process was studied on [11] by implementing a mirror effect-based learning system capable to filter user behaviors in order to extract the best ones, learn them and optimize the instructions to give to subsequent users who can cooperate or compete with the system. Cooperation and competition implementation can also be motivated by the benefits, the costs or the risks these activities may generate [41]. Human can learn from system and vice-versa but this learning process is usually built without education, i.e. explanation, justification or understanding for instance. The development of heuristics-based or rules-based approaches [27, 28, 30, 41–44] can be relevant to do this possible.

References

1. Gallez, C., Motte-Baumvol, B.: Inclusive mobility or inclusive accessibility? A European perspective. Cuadernos Europeos de Deusto, 2017, Governing Mobility in Europe: Interdisciplinary Perspectives, vol. 56, pp. 79–104. ffhalshs-01683481f (2017). Inclusive mobility: educational inclusion
2. Poltimäe, H., Rehema, M., Raun, J., Poom, A.: In search of sustainable and inclusive mobility solutions for rural areas. Eur. Transp. Res. Rev. **14**, 13 (2022). https://doi.org/10.1186/s12544-022-00536-3
3. Vanderhaegen, F.: Weak signal-oriented investigation of ethical dissonance applied to unsuccessful mobility experiences linked to human–machine interactions. Sci. Eng. Ethics **27**, 2 (2021). https://doi.org/10.1007/s11948-021-00284-y
4. Loitsch, C., Müller, K., Weber, G., Petrie, H., Stiefelhagen, R.: Digital solutions for inclusive mobility: solutions and accessible maps for indoor and outdoor mobility. In: International Conference on Computers Helping People with Special Needs, 11–15 June 2022, Milano, Italy (2022)
5. Correia de Barros, A.: Inclusive design within industry 4.0: a literature review with an exploration of the concept of complexity. Des. J. **25**(5), 849–866 (2022)
6. Culley, S.J., Hicks, B.J., McAloone, T.C., Howard, T.J., Dong, A.: Supporting inclusive product design with virtual user models at the early stages of product development. In: Proceedings of the 18th International Conference on Engineering Design (ICED 2011), Lyngby/Copenhagen, Denmark, pp. 80–90 (2011)
7. Johnson, R., Kent, S.: Designing universal access: web-applications for the elderly and disabled. Cogn. Technol. Work **9**, 209–218 (2007)
8. Singh, S., Mahanty, B., Tiwari, M.K.: Framework and modelling of inclusive manufacturing system. Int. J. Comput. Integr. Manuf. **32**(2), 105–123 (2019)
9. Vanderhaegen, F., Nelson, J., Wolff, M., Mollard, R.: From human-systems integration to human-systems inclusion for use-centred inclusive manufacturing control systems. IFAC-PapersOnLine **54**(1), 249–254 (2021)
10. Vanderhaegen, F., Jimenez, V.: Opportunities and threats of interactions between humans and cyber–physical systems – integration and inclusion approaches for CPHS. In: Annaswamy, A.M., Khargonekar, P.P., Lamnabhi-Lagarrigue, F., Spurgeon, S.K. (eds.) Cyber–Physical–Human Systems: Fundamentals and Applications, pp. 71–90. Wiley (2023)
11. Vanderhaegen, F.: Pedagogical learning supports based on human–systems inclusion applied to rail flow control. Cogn. Technol. Work **23**, 193–202 (2021)
12. Schieben, A., Wilbrink, M., Kettwich, C., Madigan, R., Louw, T., Merat, N.: Designing the interaction of automated vehicles with other traffic participants: design considerations based on human needs and expectations. Cogn. Technol. Work **21**, 69–85 (2019)
13. Cacciabue, P.C., Martinetto, M.: A user-centred approach for designing driving support systems: the case of collision avoidance. Cogn. Technol. Work **8**, 201–214 (2006)
14. Hickling, E.M., Bowie, J.E.: Applicability of human reliability assessment methods to human–computer interfaces. Cogn. Technol. Work **15**, 19–27 (2013)
15. Carsten, O., Martens, M.H.: How can humans understand their automated cars? HMI principles, problems and solutions. Cogn. Technol. Work **21**, 3–20 (2018)

16. Vanderhaegen, F.: A non-probabilistic prospective and retrospective human reliability analysis method—application to railway system. Reliab. Eng. Saf. Syst. **71**(1), 1–13 (2001)
17. Vanderhaegen, F.: Human-error-based design of barriers and analysis of their uses. Cogn. Technol. Work **12**, 133–142 (2010)
18. Schwarz, C., Gaspar, J., Brown, T.: The effect of reliability on drivers' trust and behavior in conditional automation. Cogn. Technol. Work **21**, 41–54 (2019)
19. Vanderhaegen, F.: Cooperative system organisation and task allocation: illustration of task allocation in air traffic control". Le Travail Humain **62**, 197–222 (1999)
20. Vanderhaegen, F.: Toward a model of unreliability to study error prevention supports. Interact. Comput. **11**, 575–595 (1999)
21. Navarro, J., Jomard, E., Saleur, É., Abrzrd, W., Cegarra, J.: Influence of automation mode use on selection rates and subjective assessment over time. Cogn. Technol. Work **24**, 609–624 (2022)
22. Merat, N., et al.: The "Out-of-the-Loop" concept in automated driving: proposed definition, measures and implications. Cogn. Technol. Work **21**, 87–98 (2019)
23. Weick, K.E.: Sensemaking in Organizations. Sage, Thousand Oaks (1995)
24. Jensen, E.: Sensemaking in military planning: a methodological study of command teams. Cogn. Technol. Work **11**, 103–118 (2009)
25. Albolino, S., Cook, R., O'Connor, M.: Sensemaking, safety, and cooperative work in the intensive care unit. Cogn. Technol. Work **9**, 131–137 (2007)
26. Vanderhaegen, F.: Toward a Reverse Comic Strip based approach to analyse human knowledge. IFAC Proc. Volumes **46**(15), 304–309 (2013)
27. Vanderhaegen, F.: Dissonance engineering: a new challenge to analyse risky knowledge when using a system. Int. J. Comput. Commun. Control **9**(6), 776–785 (2014)
28. Vanderhaegen, F.: A rule-based support system for dissonance discovery and control applied to car driving. Expert Syst. Appl. **65**, 361–371 (2016)
29. Vanderhaegen, F.: Towards increased systems resilience: new challenges based on dissonance control for human reliability in cyber-physical & human systems. Annu. Rev. Control. **44**, 316–322 (2017)
30. Vanderhaegen, F.: Heuristic-based method for conflict discovery of shared control between humans and autonomous systems - a driving automation case study. Robot. Auton. Syst. **146**, 103867 (2021). https://doi.org/10.1016/j.robot.2021.103867
31. Carsten, O., Vanderhaegen, F.: Situation awareness: valid or fallacious? Cogn. Technol. Work **17**, 157–158 (2015)
32. Endsley, M.R.: Toward a theory of situation awareness in dynamic systems. Hum. Factors **37**(1), 32–64 (1995)
33. Jones, D.G., Endsley, M.R.: Sources of situation awareness errors in aviation. Aviat. Space Environ. Med. **67**(6), 507–512 (1996)
34. Sneddon, A., Mearns, K., Flin, R.: Stress, fatigue, situation awareness and safety in offshore drilling crews. Saf. Sci. **56**, 88 (2013)
35. Salmon, P.M., Walker, G.H., Stanton, N.A.: Broken components versus broken systems: why it is systems not people that lose situation awareness. Cogn. Technol. Work **17**(2), 179–183 (2015)
36. Dekker, S.W.A.: The danger of losing situation awareness. Cogn. Technol. Work **17**(2), 159–161 (2017)
37. de Winter, J.C.F., Eisma, Y.B., Cabrall, C.D.D., Hancock, P.A., Stanton, N.A.: Situation awareness based on eye movements in relation to the task environment. Cogn. Technol. Work **21**(1), 99–111 (2018). https://doi.org/10.1007/s10111-018-0527-6
38. Vanderhaegen, F., Wolff, M., Mollard, R.: Non-conscious errors in the control of dynamic events synchronized with heartbeats: a new challenge for human reliability study. Saf. Sci. **129**, 104814 (2020)

39. Vanderhaegen, F., Wolff, M., Mollard, R.: Repeatable effects of synchronizing perceptual tasks with heartbeat on perception-driven situation awareness. Cogn. Syst. Res. **81**, 80–92 (2023)
40. Lu, Y., et al.: Outlook on human-centric manufacturing towards Industry 5.0. J. Manuf. Syst. **62**, 612–627 (2022)
41. Vanderhaegen, F., Jouglet, D., Piechowiak, S.: Human-reliability analysis of cooperative redundancy to support diagnosis. IEEE Trans. Reliab. **53**, 458–464 (2004)
42. Jouglet, D., Piechowiak, S., Vanderhaegen, F.: A shared workspace to support man-machine reasoning: application to cooperative distant diagnosis. Cogn. Technol. Work **5**, 127–139 (2003)
43. Vanderhaegen, F., Caulier, P.: A multi-viewpoint system to support abductive reasoning. Inf. Sci. **181**, 5349–5363 (2011)
44. Aguirre, F., Sallak, M., Vanderhaegen, F., Berdjag, D.: An evidential network approach to support uncertain multiviewpoint abductive reasoning. Inf. Sci. **253**, 110–125 (2013)
45. Vanderhaegen, F., Chalmé, S., Anceaux, F., Millot, P.: Principles of cooperation and competition – application to car driver behavior analysis. Cogn. Technol. Work **8**, 183–192 (2006)

Author Index

H. Krömker (Ed.): HCII 2024, LNCS 14733, pp. 249–250, 2024.
https://doi.org/10.1007/978-3-031-60480-5

Printed in the United States
by Baker & Taylor Publisher Services

Printed in the United States
by Baker & Taylor Publisher Services